The Business of Environmental Health & Safety Management

Guidance for managers of environmental health and safety departments to assist them in achieving excellence for themselves, their organizations and for their communities

James L. Lieberman

Library of Congress Registration Number TXU001750863

ISBN 978-0-9846825-0-8

Printed in the United States of America

FIRST EDITION

Dedication

This book is dedicated to the memory of my parents Lawrence & Evelyn Lieberman for their love, devotion and their lifelong service to their friends, family and community, and to my son, Benjamin, who collaborated with me and assisted me with the typing, layout and composition of this book.

Ben, you are the son I always hoped for and you are growing into a man that I am proud to know will carry our name in to the future.

Acknowledgments

The author would like to acknowledge the assistance of Amy Allen, controller for Array BioPharma, for her patience and instruction with methods to gather financial data and collate it from a larger accounting system and Sandi Vaughn, coordinator at Array BioPharma for help with graphing.

I would like to thank Ben Lieberman for his diligent typing, formatting, researching and organization of the text, and Gary Gautier, senior technical writer at Penta Corp., for his initial editorial review and helpful suggestions. I would also like to thank Erin Garrett, graduate student at University of Colorado for her document editing and formatting of the reference section and Christina Lefevre-Gonzalez, doctorial student at University of Colorado for her help with the final formatting of the manuscript.

I would like to thank Steve Thomas, senior principle consultant at Oracle Corp, for his review of financial concepts and examples.

Foreword

Every thinking person today realizes that the world in which we live in faces significant challenges including: environmental degradation, deforestation, global warming, political unrest, and economic upheavals. Many would say that these challenges are just too big for us to tackle.

My proposition is; what man has created both for good and bad can be changed by his efforts. I believe that if we serve our organizations as enlightened EH&S managers, we can make a difference for our organizations, our communities, and our business investors. This means starting where we are, in the present moment, and having a clear vision of where we would like to be. I hope in some small way that this book illuminates a path forward for EH&S managers to make a difference[1].

Key Concept

All games in life are scored. To seek to achieve any measure of success without keeping score is like starting a journey with no destination. You may travel far and wide, you may see many sights, but your journey will have no purpose and, therefore, will make no difference.

Business keeps score in dollars and cents and measures objective outcomes. If you want to make a significant impact at your organization, you will need to clearly propose objectives, means to achieve them, and scales to measure your progress towards them. Learn to quantify wherever possible.

Table of Contents

Chapter 1

Introduction

This book will present a step-by-step approach to present the concepts and guidelines needed to integrate business concepts into EH&S management and to assist a manger to communicate the value of their department and proposals. We will use a hypothetical situation where the manager has no predecessor and is tasked with starting from scratch. Although this scenario is unlikely for most EH&S managers, it will serve us as a good starting point to illuminate the larger picture. Use this book as a reference. Choose those sections that assist you in making better and more substantiated decisions.

A Challenge for the New EH&S Manager

Imagine that you are scheduled to make a presentation to senior management of your company. You are a new manager of your company's Environmental Health & Safety Department and want to make a good impression. You have twelve months to prepare for this meeting.

What topics will you cover? What format will your presentation take? How long will your meeting last? What questions will you ask? What questions are you prepared to answer? This book will help you to prepare for this meeting and to manage your department effectively.

What do you want to convey in your presentation?

1. You understand executive management's goals and are striving to align your department to execute them.

2. You have studied your company's operation and the scope of business, you understand the body of laws and regulations applicable to the company, and you recognize the best practices to serve your company's particular interests.

3. You are managing your staff effectively. You have:
    a. Determined staffing requirements
    b. Hired or are in the process of hiring capable staff
    c. Established a system to supervise direct reports to maximize their effectiveness
    d. Set your employees annual goals and objectives
    e. Delegated required tasks to staff
    f. Planned and implemented procedures and systems to maximize staff efficiency
    g. Instituted an EH&S management system to ensure task completion

4. You are managing your resources effectively. You have:
    a. Employed financial and budget tools to maximize operational efficiency
    b. Facilitated the preparation of appropriate management reports

c. Reviewed performance data and reports to monitor and measure productivity

One of the objectives of this book is to help you to understand management's goals and speak their language, to show that you are using your staff effectively and that you are managing your resources effectively. Since there are excellent resources for advising managers on the supervision of employees, we will present our perspective and give the reader resources to expand their knowledge, but will not dwell on this topic. Our goal will be to go into significantly more detail in managing resources effectively and sustainably, and describing how to create compelling business presentations.

It has always been important to be in communication with management. It is especially difficult in this age of instant messaging to decide what information to present. Effective managers know that the majority of the information they collect should be objective and tell a story. This guide will endeavor to mentor a younger manager to prepare him/her to plan effectively to manage their department by giving them tried and true business tools and examples of the use of the tools.

A new manager needs to understand executive management's perspective. Executive management must decide where to use the company's limited resources in the most advantageous way. Usually that means selecting investments that have a high rate of return, or a short payback period. Executive management is well aware, especially in young companies, that capital is a precious commodity. When managers ask for resources, they must understand that their request will be ranked against others. If a new EH&S manager cannot make a compelling business proposal, then the likelihood of their projects being approved will be significantly impaired.

The second part of the guide will be directed towards coaching managers in how to prepare for one of their most important tasks: briefing executive management on the efforts, accomplishments, status, and future goals of their department. Experienced managers know they don't get a second chance to make a first impression. For better or worse, the impression the new manager makes will stick with them, so it is incumbent upon the manager to do the very best job they can the first time around.

This will involve a significant amount of work; I will be your coach, and you my athlete. I will be endeavoring to provide insight and wisdom to make your efforts productive, but you will need to provide the effort and focus to get in shape to compete. Make no mistake, I mean compete. You will compete for executive management's time and for corporate resources, and you will be compared to your peers. To win, you need to know how the game is played and how the score is kept.

Chapter 2

Management Philosophy

Every EH&S manager will have an underlying philosophy for their approach to environmental health and safety management. In this chapter, I lay out my ideas gleaned from my training, education, and experience. Use them to examine your own philosophy. The clearer you can present your concept of how to conduct the business of EH&S management, the better prepared you will be to answer more difficult questions.

Philosophical Underpinnings

This book will argue that an EH&S manager should strive to present as much information in an objective format as possible. All committed managers have an overarching philosophy that needs to be shared with their staff and management. When the time comes to translate that vision into a specific plan of action, every effort should be made to present available information in an objective format.

Objective data should be organized to bring out the information that can help to illuminate important business decisions. Every day, we need to make decisions. We need to empower our staff to make decisions. We need to communicate with executive management and customers. If we cannot demonstrate added value for a particular suggestion or proposition, then either (1) it wasn't a good idea or suggestion to begin with, or (2) we didn't do the necessary work to get the facts together coherently and demonstrate objectively that the idea would add value.

Does this mean we constantly are reducing all of our suggestions to dollars and cents or something numerically quantifiable? Of course not, but those objectives we are willing to push as larger organizational goals will be vastly more convincing if they yield quantifiable benefits.

As I make the postulation that we should demonstrate value in objective terms, I can hear the chorus of objections from my fellow EH&S managers. To them, my argument would be as follows: Our job involves developing human attitudes and motivations, risk avoidance, interpersonal communication, and sustainability in addition to more tangible tasks such as compliance, risk management, best practices, and business continuity.

Many would argue that these tasks are by their very nature not quantifiable. How do you quantify an attitude? How do you quantify the avoidance of a negative outcome? How do you quantify job satisfaction? How do you quantify that your company is pursuing the triple bottom line[2], which measures business success not only by the traditional bottom line of financial performance, but also by their impact on the broader economy, the environment, and the society in which they operate.

I certainly agree that our work needs to deal with such intangibles as attitude, motivation, job satisfaction, and the like, but I would argue that all of these things have quantifiable impacts in the company. EH&S managers who succeed as integral players in the company structure will be those who make the best effort to be objective and to quantify information.

Will it be easy? No. Will it be fulfilling? Yes. It is always better to light one candle than curse the dark. This book's goal will be to assist you, the EH&S manager in your effort to light candles. I hope to demonstrate that one may look at what appears at first to be completely subjective information for decisions and reframe the discussion so as to make the subjective, objective.

Let's start with an example of something as soft, but as we know critical, as employee attitude. How would you measure it? Well first, you would want to reframe the question and ask: What attitudes do my fellow employees, line managers, or executive managers have towards X? Now let's substitute a specific question for X, such as "Do you feel that our company conducts sufficient safety inspections of our laboratories?" You would want the answer scale to be numeric. For example, 1 would be strongly agree, 2 would be agree, 3 would be neutral, 4 would be somewhat disagree, and 5 strongly disagree.

Obviously asking one question would not get you to the full objective of understanding the general attitude towards laboratory safety. However, if you craft a survey well with an objective scale, you will be able to get numerical output that will be comparable to a future survey incorporating the same or a similar question.

Implicit in this discussion is that enlightened EH&S managers are not satisfied with the status quo, but want to work for continuous improvement. If your survey is well crafted, it will illuminate areas where your employees, managers, or executives feel that you are doing things safely, environmentally consciously, sustainably, and so forth, as well as areas where you could be doing better. What you now have is a value in a point of time.

As we know, point source data yields little information about the status of a larger system. But we also know that a series of point source measurements over a period of time will yield valuable statistics about the system. For instance, if the answer to the question posed averaged as a 4, you would likely decide that your staff is not conducting sufficient lab inspections *to create an attitude in the mind* of the survey group that there were sufficient lab safety inspections.

This value conclusion would be completely independent of the compliance requirement. Perhaps your department meets every federal, state, local, and corporate requirement for the frequency of lab inspections. Nevertheless, you would not have created the attitude *in the mind* of *the surveyed group*, that sufficient inspections were being made. Therein lays the opportunity for you to make a management decision. Do you increase the frequency of laboratory inspections? Do you meet with laboratory personnel to find out what may underlie the attitude? If the question was worth asking, then the follow-up will be worth pursuing.

Here are some tools for creating and conducting surveys:

The Survey Systems[3]
SurveyMonkey[4]

Over time, by applying a tool to measure attitudes towards specific facets of your EH&S program, you will gain knowledge that will empower you to make better decisions. This is one of the essences of good EH&S management. The goal of this book will be to empower new and seasoned EH&S managers to effectively craft tools to objectively demonstrate to you, your staff, company, employees, and senior management that your programs are making a difference. Whenever possible, we will measure the difference in dollars and cents. If the question is worth pursuing, we will always measure the answer as objectively as possible.

During my career, I have worked in a university laboratory, for a resort management company, as a consultant to Department of Energy Facilities, as the president of my own consulting firm, and I have had the pleasure of serving for nine years as the manager of EH&S for Array BioPharma, a drug discovery company located in Boulder, Colorado. I have had the chance to see first-hand how many different companies approach the question of how to comply with a vast and growing body of environmental health and safety laws and regulations.

During my tenure, the world has changed from more of a laissez-faire attitude towards EH&S to one that realizes that EH&S concerns are important for employees and the bottom line, and should be given appropriate attention.

In the past I found that in most organizations, EH&S concerns take a back seat to other "more pressing" concerns faced by management. I believe this to be so in part because we, EH&S managers, were not able to present the value in the service that we provide. Unfortunately, until recently, we have not focused on value added for our organizations, but more on compliance.

Some organizations have made what I believe are altruistic but contradictory statements, such as "safety is number one" or "our goal is an accident free work place." As noble as these slogans are, I think they create a paradox. That is, we are not really telling the truth. EH&S is not the most pressing concern of our corporation. The primary concern of a corporation is to make a profit. We do ourselves and our profession a disservice by creating a false goal to pursue. In all conscious relationships, the basis for success is telling the truth, and the truth for all business organizations is that they must succeed in their primary mission, that being paying their employees, management, suppliers and producing a profit to share with stockholders. Anyone who disagrees with this postulate, I believe, is ignoring the company's big picture, and will be at a disadvantage when trying to articulate the proper value of EH&S to the company as a whole.

Unfortunately, in the twentieth century, we have seen false but great sounding mottos, such as *The Communist Manifesto's* "From each according to his ability, to each according to his needs," which have spawned horribly oppressive governments and regimes. Fortunately in the United States, our forefathers grew up in a period of enlightenment, when reason was applied to political and moral questions. The best governments were ones that realized that in economic terms Adam Smith's idea, that we will act in pursuit of our own self-interest, describes in fact the true nature of human kind. We are self interested.

No amount of hand waving, sloganeering, smoke and mirrors, praying, or meditation will change the fact that we are creatures of self-interest, so enlightened EH&S managers will not try to go against human nature. Rather, they try to align themselves with it. One of the best approaches to safety is expressed by Scott Geller in People Based Safety.[5] One of his central ideas is that EH&S managers should create an atmosphere where employees do things safely because *they realize that it is in their best interest*. I believe that his method of inculcating a positive attitude and concept towards safety is the right one.

So let us be honest: EH&S is one of our employer's concerns and it is not necessarily the paramount one. But skilled EH&S managers will endeavor to recognize this truth and use it to their advantage. The policy that I set forth for Array BioPharma was not a zero tolerance for accidents, which I believe may have fostered a fear of accident/incident reporting by employees, but rather that we should work together to reduce accidents and injuries to the lowest level reasonably achievable. That was a goal I felt comfortable promoting and pursuing. The EH&S manager will need to build sustainability compassionately and rationally within and throughout their organization.

Chapter 3

Visualizing EH&S Management

Sometimes a picture is worth a thousand words. In this chapter, I will use a pyramid illustration to represent Abraham Maslow's hierarchy of needs, which focuses on the stages of human growth. We can use Maslow's ideas as an analogy for the development of a sustainable EH&S program. I believe the value of this analogy is that it allows you to communicate a vision to executive management, staff, and employees to describe where your department and company currently lies on the evolutionary path of sustainable EH&S.

Let's start of by reviewing Maslow's hierarchy of needs[6], and then see how these descriptions of needs may be applied analogously to sustainable EH&S management.

Abraham Maslow, in his 1943 paper "A Theory of Human Motivation"[7], described a hierarchy of human needs. His theory is based on his observations of humans' innate needs and curiosities. It focuses on describing the stages of growth in humans.

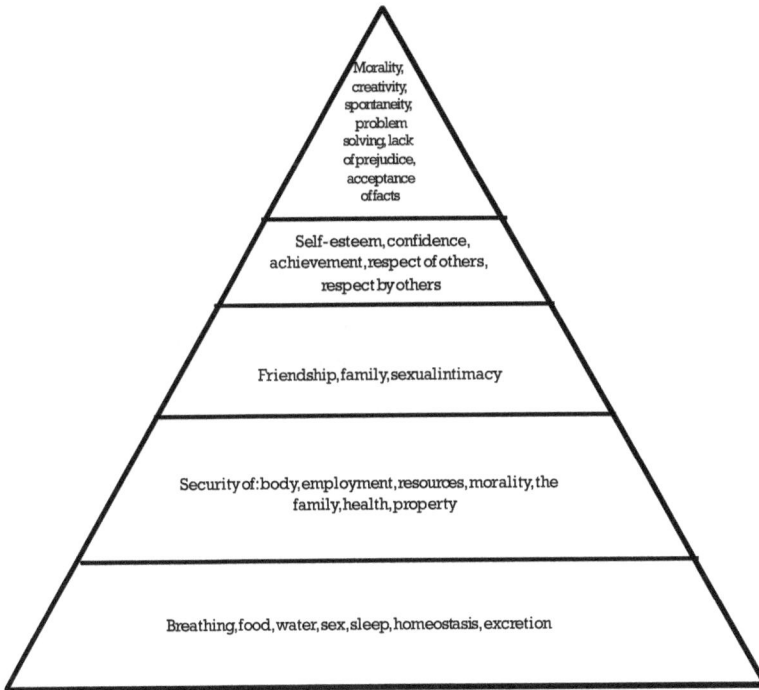

Morality,
creativity,
spontaneity,
problem
solving, lack
of prejudice,
acceptance
of facts

Self-esteem, confidence,
achievement, respect of others,
respect by others

Friendship, family, sexual intimacy

Security of: body, employment, resources, morality, the
family, health, property

Breathing, food, water, sex, sleep, homeostasis, excretion

The base of the pyramid includes physiological needs

Physiological needs are the requirements for human survival. If these requirements are not met, the human body simply cannot continue to function.

Air, water, and food are metabolic requirements for survival in all animals. Clothing and shelter provide necessary protection from the elements. The intensity of the human sexual instinct must be adequate to maintain a birth rate conducive to survival of the species.

Safety needs were next

Once a person's physical needs were relatively satisfied, the individual's safety needs would take precedence and dominate behavior. These needs have to do with people's yearning for a predictable, orderly world in which perceived unfairness and inconsistency are under control: the familiar frequent and the unfamiliar rare. In the world of work, these safety needs manifest themselves in such things as a preference for job security, grievance procedures for protecting the individual from unilateral authority, savings accounts, insurance policies, reasonable disability accommodations, and the like.

Love and belonging

After physiological and safety needs are fulfilled, the third layers of human needs are social and involve feelings of belongingness. This aspect of Maslow's hierarchy involves emotionally based relationships in general, such as friendship, intimacy and family.

He believed that humans need to feel a sense of belonging and acceptance, whether it comes from a large social group, a tribe, religious groups, professional organizations, sports teams, or small social connections. He believed that humans needed to love and be loved (sexually and non-sexually) by others. In the absence of these elements, the thought is that many people would become susceptible to loneliness, social anxiety, and clinical depression.

Esteem

He believed that humans have a need to be respected and to have self-esteem and self-respect. Also known as the belonging need, esteem presents the normal human desire to be accepted and valued by others. People need to engage themselves to gain recognition and to have an activity or activities that give them a sense of contribution; to feel accepted and self-valued, be it in a profession or hobby.

Maslow noted two versions of esteem needs: a lower one and a higher one. The lower one is the need for the respect of others, the need for status, recognition, fame, prestige, and attention. The higher one is the need for self-respect; the need for strength, competence, mastery, self-confidence, independence and freedom. The latter one ranks higher because it rests more on inner competence won through experience. Deprivation of these needs can lead to an inferiority complex, weakness and helplessness.

Self-actualization

This level of need pertains to individuals fulfilling their potential. Maslow describes this desire as the desire to evolve into what one really is, to become fully developed. This is a broad definition of the need for self-actualization, but when applied to individuals it is specific. For example, one individual may have the strong desire to become a world class scientist, in another the need may be expressed artistically, and in another it may be expressed athletically. The individual must first achieve the previous needs: physiological, safety, love, and esteem, before tackling self-actualization.

Visualizing Sustainable EH&S

Let us create an analogy to Maslow's hierarchy to describe the hierarchy of EH&S needs for a modern organization.

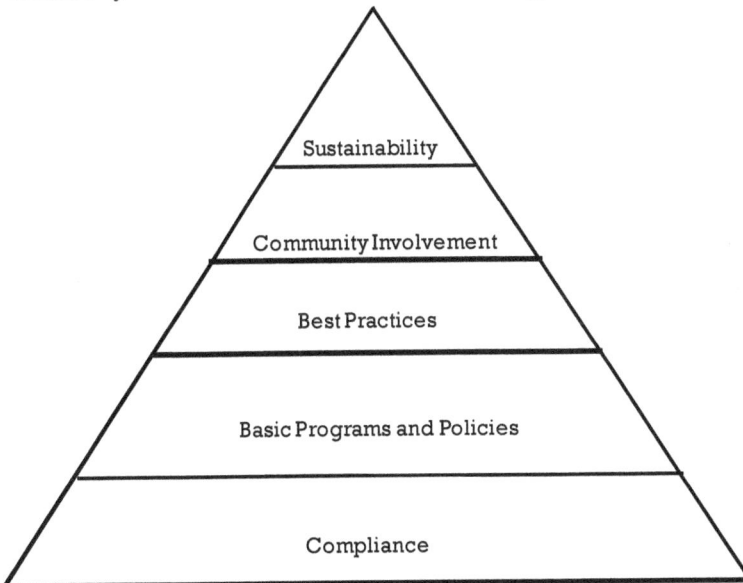

Compliance

At the base of our triangle we would place core compliance. Every corporation that wants to continue to exist needs to comply with the basic laws and regulations of the country it operates within. This would be analogous to Maslow's physiological needs.

Basic Programs and Policies

As we proceed up the triangle, the next layer is basic programs and policies to implement the compliance requirements. This would be analogous to Maslow's safety needs, in that we need to create a reasonable, predictable, and consistent set of guidelines for our company's employees to follow.

Best Practices

The next level up would be the institution of best practices. Best practices are policies, programs, and methods to achieve excellent EH&S programs that are specific to certain industry groups. For maximum results, best practices need to be integrated into your company's normal operations so that EH&S requirements become indistinguishable from normal corporate practices.

Community Involvement

The next level of hierarchy is the company's integration into the community. Every EH&S department needs to interface with outside groups such as: fire departments, state regulatory bodies (such as the state Department of Public Health and Environment for requirements surrounding air, water, and waste management), and other compliance areas. The conscientious EH&S manager will reach out to other entities such as the local chamber of commerce, the local department of health, possibly a local HAZMAT team, and the local emergency planning committee.

Sustainability

The pinnacle of the EH&S hierarchy is sustainability. At this point, the company will be exceeding the normal EH&S requirements. It will be using its knowledge and abilities to enrich the community. The company and the EH&S department has matured to a state in which substantially all other requirements have been met and attention can be turned towards achieving the triple bottom line[8] of continued economic success, while protecting the environment and improving the lives of whom the organization interacts with.

Now that we have lain out the visual map needed to conceptualize the journey that our organizations would need to travel to reach sustainability, let's address some of the tasks necessary to reach the goal.

1. The company needs to be financially successful. That is, it has to have sufficient ongoing revenues to enable it to support the formation of an EH&S department, and be able to plan for at least a three year time horizon.

2. The company needs to have a functioning EH&S department with a competent manager, and enough resources to carry out the mission.

3. The EH&S department needs to understand the business's operations, and have researched applicable laws and regulations in sufficient detail so as to have identified those requirements that the corporation must meet. It has instituted an EH&S management system to ensure that necessary tasks to meet the requirements are in place.

4. The EH&S department's programs, policies, and guidelines must be integrated into the normal operation of the company. The manager has reached out to other similar businesses in a non-competitive environment to discover and share what best practices are applicable for their organizations.

5. The EH&S manager has looked into the community in which the corporation resides and has determined that the company now has the resources to proactively reach out to regulatory

and community groups such as the fire department, the chamber of commerce, etc., to begin to establish personal relationships. Furthermore, the company has determined that its organization's knowledge and skills can make a positive contribution to the community by being a leader or supporter of selected community organizations.

6. The EH&S manager with his department and corporate support will need to develop objective measures of performance in the areas of profitability (cost containment), environmental protection, and community support that go beyond traditional compliance requirements.

The rest of the book will strive to describe in sufficient detail how EH&S managers may proceed from step to step to bring their organizations to the pinnacle of sustainability.

Let's start from the beginning: with you. If you seek to be a sustainable EH&S manager, then you realize that you must lead by example. For instance, if you seek to have an employee wellness program, you must either be committed personally to taking care of your health, or be willing to acknowledge that you are not living healthy and will yourself seek to change and participate in the program that you are proposing. If you fail to do either, you will be out of integrity. Whenever you are out integrity you significantly reduce your own power to make a difference in the world.

Please understand that I am not proposing that you be perfect before you try to manage an EH&S department. That would of course be ludicrous, but in order to be a visionary EH&S manager, you must be willing to grow personally and professionally. Your perspective will determine your response to this challenge.

A superior EH&S manager is inner directed. You must be in alignment with your goals. The payoff for being in alignment with your goals is a sense of purpose, achievement, and vitality. So to be a superior EH&S manager, you must honestly seek to apply what you learn to inspire yourself, your family, your co-workers, your community, and, in effect, the world.

This looks like a huge challenge if you look at it in its entirety, but using a different perspective of incremental change, it is not.

The longest journey starts with the first step. You should measure success; this book will encourage you to measure success, not from a utopian goal backwards, but from the reality of exactly where you are now forwards. The task is a continuum. Hopefully you will carry the baton forward until you can pass it off to the next manager. In the words of Elie Wiesel[9],

> But I know this: the questions that confront us today do have a response; and the response engages us. If the present world has a purpose or fate, it must be the same for all. And each human being, with his own background and culture, owes it to him/herself to affirm his/her own humanity with respect to that of his/her peer. The purpose of this world cannot be to propose or impose a choice between joy for some, and distress for others. This is a false or unjust choice. If, in order to be happy, it is necessary for others not to be, the world in which we live would look more like a prison than an orchard.

> Transforming the whole world into a massive enclosure, is indeed the goal of a fanatic suffering from ugly and unappeased hatred, not of a sincere and warm-hearted believer. The former—the jailer—aspires to stifle out all those who are not like him. The truth is that he manages to put God himself in prison.

> *Man's task is thus to liberate God while freeing the forces of generosity* in a world teetering more and more between curse and promise...though God created the world, *it is up to people to preserve, respect, enrich, embellish, and populate it without bringing violence to it.* (Italics added)

So if I haven't scared you off by now, I hope I have set the basic philosophic outline of this book. With the philosophical groundwork laid, we can move on to the more concrete task of how to achieve EH&S sustainability.

Visualizing Our Relationships

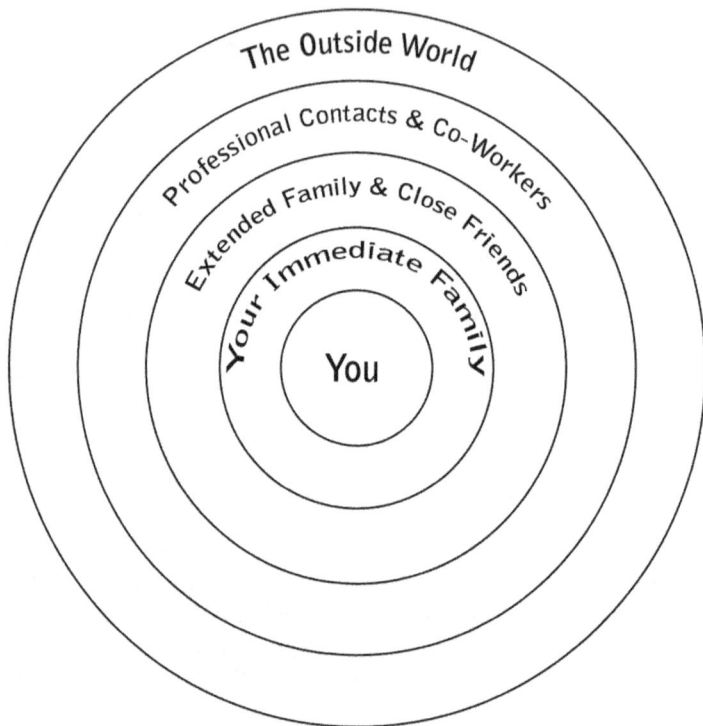

I would like to visualize what I believe is a model of our sphere of influence. Again, it starts with you in the middle.

The next circle is your immediate family. You have the most influence with and your actions make the biggest difference to those closest to you in your concentric circles of relatedness. The next circle would be your extended family and close friends, followed by your professional contacts, co-workers and, finally, the outside world.

You owe the highest responsibility to your inner-most circle, followed by lesser degrees of responsibility to those in the more distant circles from your center. It will be up to you to balance this effort, but in doing so you will apply your greatest wisdom. There is an old Talmudic saying that goes, "If I am not for me, who will be for me, and if I am only for myself, what am I?" Bob Conway, the CEO of Array BioPharma, once said to company managers that if they did the right thing for the company's employees, they would be looking out for the company's best interest.

I believe him. In the corporate world, your circles project outward from your department. Obviously, you will need to make sure that you do the right thing for your reports. If you do not take their interests, needs, and desires into consideration, and assist them in reaching their professional goals, you cannot expect them to provide the level of support and faithfulness you will need as a basis to disseminate information, complete tasks, and model the type of concern you want to project.

Next, of course, are the company's employees. They will quickly pick up on the atmosphere that your department creates. If you set the tone that your department is a service center, and you express your department's goal to provide a high level of service and support to employees to assist them with EH&S matters, your department will be look upon as a resource, not as a policeman.

Remember that your goal is to create a willingness and active acceptance within your company's employees to comply with your interpretations of regulations and good practices. If you and your department take the time to explain the basis of why employees need to do or not do some action, you will get their buy-in. If you solicit their input and use it, you will lay the foundation of mutual trust. This level of trust is a key component of your effort.

My teenage son has taught me a lot about management. Teenagers have a propensity to make poor decisions, as we all do on occasion. The enlightened parent seeks not to punish the child (although sometimes a punishment is appropriate), but for the child to learn from their mistakes. Therefore, your goal as a parent and manager should be to point out the mistake in a way in which your child/employee can see the truth of it for themselves. This is the way you teach inner directed individuals. Then your task will be to provide them with an opportunity to learn the lesson from the mistake so that they can grow and avoid similar mistakes in the future.

Philosophically, you will have an easier time with outwardly directed people, those that require constant outward direction with reward and punishment to comply, but in practice such individuals are challenging to direct.

Most organizations recognize that they want to minimize the number of outwardly directed types of employees in the organization, because they will need a constant carrot and stick to comply. Hopefully, you will recognize that the vast numbers of employees in your organization are inner directed, and as such you appeal to their heads and hearts first. I strongly suggest that all EH&S managers read and try to apply the concepts described in the book, <u>People Based Safety</u>[10].

Chapter 4

Understanding Executive Management

One of the first questions a perspective speaker should ask themselves is: who is the audience? You need to tailor your conversation and graphical presentation to fit the knowledge, experience, and pre-conceived ideas of your audience. You need to understand that there are "Tribal Languages". As EH&S managers, we speak with an alphabet soup of acronyms such as: OSHA, PEL, OEL, RCRA, TSCA, etc. One of the quickest ways to lose your audience is to talk in a "different language". If you want to be effective with executive management, you will need to speak their language. This chapter will assist you in identifying executives' concerns and the language of finance.

Businesses Run on Money

Profitability is indispensable to business success (except for a very early stage company, where basic survival is paramount). A lack of profitability shows poor management. Therefore, profit is the normal basis for business decisions and the customary gauge of business performance.

In order for a business to operate, it must have a sufficient revenue stream to cover such expenses as employees' wages, rent/mortgage/taxes, suppliers, and the purchase of other required assets. The most important thing to know about capital and business is that money must be coming in faster than it is going out. In other words, the company must make a profit. Without making a profit, there will be a loss of investors, capital, and eventually employees.

Knowledge of finance is essential for almost every management position. Events, rules, problems, and decisions will almost always involve finance. Executive financial decisions affect you, your position, and your entire company. Therefore it is important to understand what current financial events are occurring and understand the way businesses work.

At the department level, your written reports submitted to your supervisors should include financial measurements. For example, the expenses this quarter as compared to the expenses generated last quarter or the expenses budgeted. Comparison is an important part of your reports.

If you are proposing a project or idea, expected profitability, a reduction in cost, or an increase in efficiency is required to make an effective argument.

Executive management's job is to make sure that your business/firm has sufficient liquid capital to pay for current expenses, and that the business is not undercapitalized (lacks sufficient investment funds). Therefore they will rank most projects by the project's capital needs and the rate of return.

Profit can be measured in many different ways. One way is called the profit margin: how many cents a business keeps out of every dollar in sales. Profit can also be stated as a total amount of money, but without being compared to a standard of reference this is of little value. It can also be expressed as a percentage of resources used to make the profit. This is a more informative and useful way to present your profit because it reveals the effectiveness of the use of your assets/resources.

Departments that do not contribute to product or service creation or sales (therefore not directly to profitability) must show that they are controlling costs. We refer to departments that do not directly contribute to profitability as administrative or "overhead" departments.

In almost every case executive management (except for cost plus contracting) tries to reduce overhead costs to a minimum. Although it is difficult to reduce costs in an administrative department, executive management expects department heads to keep the annual cost increases to a minimum. Administrative managers must understand the executive manager's expectations and be able to demonstrate cost control efforts objectively.

Chapter 5

Management and Communication

Being an EH&S manager requires the individual be in communication with executive management, their staff, and employees. Although the manner in which the communication is delivered will vary with the audience, the underlying goal should be constant, that being creating a safe and supportive work place that is in compliance with regulations and good practices in an efficient and economic manner. The manager needs to enroll many other players to realize the goal. To do this, they must be cognizant of others' motivation and needs. This chapter highlights management of a department and communicating with EH&S management and staff.

A manager's principle task is to improve performance within the department. Many obstacles stand in the way of this goal. I know that you want to be a great manager. To do so, you must be someone who directs a team to solve problems. You must manage both people and resources effectively.

Setting Department Goals

When enlightened managers discuss objectives with their employees, they must be sure that they are plausible and that they further the goals and objectives of the department and company. After employee goals have been set, the time frame for completion should be discussed and finalized. After you have given your employees their assignments, you should give them as wide a berth as possible to get their work done. As long as their final result satisfies your expectations, there is no reason to micromanage them.

Micromanaging results in losing the trust of loyal employees, and setting the manager himself on an exasperating path. If the assignment is large, you should set milestones to be completed and meetings to discuss the project's status. This will ensure that your employees are working up to your standards and that the final product will be what you expect it to be.

Outstanding managers and supervisors do more than direct — they lead, build morale, encourage teamwork, and create enthusiasm.

As a manager, what kind of relationship would you like to create with your employees? What types of relationships do you believe would be the most productive and satisfying for you both?

Leader? Friend? Teacher? Mentor? Other?

Forward-thinking managers have discovered that the same skills coaches use to create winners in athletics also work in a business setting. Coaches motivate their athletes, guide them, encourage them, and help them to establish a winning attitude. The benefits of a manager/coach are:

- Transforming your work group into a cohesive, coordinated team
- Setting your employees on track for success
- Recognizing and supporting exceptional employees to help them realize their fullest potential

Find out what motivates your employees and, if at all possible, make sure you reward them with it, whether it be:

- More money
- A new title
- A flexible schedule
- Expanded work in their area of interest
- Promotion

Remember a coach is not a peer; you can be friendly with your staff, but is very difficult to be a friend and a supervisor. A supervisor needs some detachment. At work, befriend other managers and employees outside of your area. For your reports, be a leader, teacher and mentor. You will find mentoring to be especially rewarding.

In order to learn to become a great manager and to achieve the results you want, you will need to:

- Learn basic management skills
- Communicate effectively
- Delegate responsibility
- Motivate employees
- Build a strong team

Managing Your Team

Select people to join your department based on ability, desire, discipline, personal responsibility, and whether or not they're a team player. To develop the individuals that you have selected, find the right set of tasks to assign them.

Make sure that your employees know that each is a team player and that the success of the team will, in part, determine their personal success. You can do this by describing the guardian angel approach.

That is, each employee in your department has a responsibility to look out for their co-workers' best interests. If they see a drop out, it is their responsibility to have a constructive conversation with their co-workers to help them insert the correction.

It is *NOT OK* to allow your co-worker to fail if you easily see the solution that is in his/her blind spot. Let your employees know you expect them to go beyond cooperation and to actively assist each other. Let your employees know that they can always confide in you with absolute certainty of confidentiality so that they will seek out your advice. *Make sure you never chastise an employee for asking for advice, especially in sensitive areas. (Whenever an employee has a concern about a co-worker is a sensitive area.)*

Many great managers have agreed on one thing: rather than focusing on people's weaknesses, you should develop an employee's natural talents. Following this path will allow the employee to feels more satisfied and empowered to perform the work you assign them. They will also feel more grateful towards their manager (you) for providing them with work they'll enjoy doing and executing completely.

The diversity of a manager's group of employees is an extra benefit because each employee has unique strengths and weaknesses; one individual's strength can help to make up for a weakness in his co-worker. Make sure that employees work together so that their corresponding strengths offset others' weaknesses.

For example, if you hire a person with great people skills but a less developed knowledge of regulations for your department, you might place them in a group with a less socially adept but very knowledgeable co-worker; together they can find a synergy to complete assigned tasks.

This does not mean that managers should ignore the weaknesses of their employees. Part of an employee's performance objectives should require them to constructively address their weaknesses. As a good manager you should provide coaching, training, and opportunities for them to improve themselves, but their weakness should not be your focus.

Finally, a great manager is not afraid to share department performance measurements, accomplishments, and goals with their employees. In fact, a great manager looks forward to giving his employees the big picture because it is a powerful motivator. When employees feel like they are being included in the big picture, they feel like a member of the team. Be sure to always publicly praise your staff's accomplishments, whether it is in a department setting or a larger corporate setting.

Relevant information you have collected should be shared with your reports. Remember not to share salary or individual employee performance information in any group setting. *Be very careful about setting one employee up to compete with another. This is a recipe for disaster.*

Insuring Team Members' Execution

A future chapter will explore EH&S Management Systems. One of the best ways to ensure that your department members execute required tasks on time is by clearly listing and delineating recurring tasks that your staff needs to complete. By systemizing the presentation and documenting the completion of the recurring task, you empower your staff to plan their own schedule to complete tasks. This will free you from day to day worries concerning the completion of required compliance tasks and will let you focus on developing new or expanded programs.

This will also present you with an opportunity to engage your staff in discussion about current challenges: Use it. I suggest meeting as a group at least monthly to discuss the completion of non-recurring tasks, new challenges that your staff members would like to pursue, and follow-up reports on delegated special projects. One of your jobs as a manager is to keep your staff engaged. You do that by recognizing their talents and selecting new projects for them to tackle. Use your monthly meeting or set up a special meeting to discuss the progress that your staff members are making on new projects. Stay in communication; that is the best way to ensure staff execution.

Communicating With Your Immediate Supervisor

You as a manager have a duty and responsibility to keep your immediate supervision appraised of your department's objectives and goals, and how you are progressing towards them. Your supervisor's time is a precious commodity. Try to organize your department's progress reports in a way that empowers your boss to communicate with his peers intelligently on your department's behalf.

Below is an example of how to succinctly communicate a department manager's efforts and successes to a supervisor. The example is a written annual report, but you should meet face to face at least once per month and informally discuss matters. Note that you should always consult with your supervisor before embarking on any project that may have political implications. Use his or her knowledge and experience to help guide you and help clarify what is and is not possible within your organization. If you cannot get your supervisor's buy-in, then you should not pursue the project. Never embarrass your supervisor by allowing him to be blindsided concerning some action in which your department may be engaged. He should be the first to know and comment on the acceptability of any significant new effort.

July 2003 to June 2004
Major Accomplishments
Summary

1) Investigated problems and determined a permanent fix for weighing hood performance. Created an installation design specification to make sure that future hood installations would meet performance requirements.

2) Created disposal guidelines for liquid non-decay radioactive materials. Our company has expanded its use of radioactive materials, so we had scintillation, digestion and plasma liquids to handle. I did not want to create a mixed waste, so I worked out a solution with the scientist. Result: Reduced disposal cost and increased convenience for scientist.

3) After I was informed that we had some old isopropyl ether, I upgraded EH&S Shock Sensitive Material guidance. Coordinated a companywide search for materials of concern and arranged for the material disposal.

4) After continuous problems with the correct use and maintenance of the hydrogenation equipment, I met with chemical lab managers and mapped out a plan for the modernization of the hydrogenation equipment and layout, improvement of the rooms housekeeping, and training of personnel. Significantly, we have reduced the number of accidents and improved the equipment maintenance for both locations.

5) I have almost completed research for a serum banking provider. The current vendor FBS increased our basic fee about 20% per year for the last few years. I hope to reduce the price we pay from $500 per month to around $50 per month. The biggest problem is that we have small quantities. I also wanted to identify a vendor that may prove useful to clinical. I am convinced we will want to store clinical or redundant samples off-site in the future.

6) Updated our volunteer bio-sample collections to include an in-house phlebotomist and updated policy for use of in-house phlebotomist. Results: We have three extra phlebotomists for flexibility and scheduling, and it is clear that there is a distinction between R&D and occ. med. sample collection.

7) Completed a very challenging return-to-work program for JC. Results: When he left the company there were no EH&S complications.

8) Worked with Process Chemistry, DH and TE, on safety for large scale cyanide and hydrofluoric acid reaction to ensure that we have reduced the hazard to minimum and had planned contingencies. The reaction went off without a problem.

9) Reviewed the shipping personnel and vendors to see if we could reduce the cost of chemical transportation and ensure flexibility. The result was that we added FedEx Ground transportation, and we will have 2 more general shippers and 2 more shippers limited to biological and R&D samples. As the company grew and we added translational, our need to send samples out increased and necessitated more trained people.

10) Responded to a laboratory indoor air quality complaint. The complaint required fairly extensive investigation to determine the cause and to make a recommendation for a fix. The result was happy scientists and a concrete recommendation to facilities for a fix to the ventilation system at the far east end of 260.

11) Created dioxane/ammonia and other liquid/gas mix storage guidance. This was necessary because in analyzing a near miss we determined there was a significant hazard. The problem was caused by freezing the dioxane mixtures. When thawed, containers exploded. The guideline should result in the elimination of this hazard.

12) MSDS for API & R&D compounds: Hopefully, the web location of MSDSs and added categorization will facilitate the dissemination of safety information, and make compliance with the law as easy as possible for the scientist.

13) Compliance programs for ANC: Gathering information to put basic compliance programs in place. Although this was not necessarily that difficult, it did require coordinating with a number of people, delegating tasks and trying to make good decisions for a distant location. We should have this complete in mid September.

14) Worked with VS and HM to create an all–company update to clarify injury reporting and API categorization. Impetus was to avoid employee confusion about return to work requirements, especially for non-occupational injuries. This was extremely important for a return to work issue in July/August. It is important not to encumber HR nor EH&S in taking proper actions.

15) Task delegation for VS, PM, and HM. This is one of my most challenging tasks. I have created and continue to update recurring tasks for my staff. I want to keep them busy, but not overwhelmed. This can be challenging. HM needs a very flexible schedule because of her family demands. One week she may work 25 hours, and the next she can only work 10. Thus I delegate to her more difficult, distinct tasks, such as the biennial waste report; something with a finite scope and a deadline. PM is growing so I need to add to his list of duties and have, but I need to insure that he has the time and

flexibility to complete his primary hazardous waste disposal task.

VS tasks are growing, because the company is growing. I need to take something off of her plate; I have given PM some of her tasks, but the challenge is to work smarter not harder. That is why I have been pursuing a learning management and recurring events system for over a year. It has been very frustrating because we have come close but every time there has been an IT problem. I have L, one of VS's friends, working on an adaptation of master control, software we are using in GMP. I am hopeful that we can use it. If so, it will take a significant burden off of VS.

I hope that this example of a comprehensive but succinct communication to supervisor will aid you in crafting your own.

Additional References:

This book has given some advice on the management of human resources, but it is not my emphasis. For further reading on the subject, I suggest the following books:

> *Built to Last,* Jim Collins and Jerry Porras[11]
>
> *FYI For Your Improvement,* Michael Lombardo and Robert Eichinger[12]
>
> *The Leadership Challenge,* James Kouzes and Barry Posner[13]
> *Good to Great*[14]

You can also attend one of the American Management Association[15] or Pryor trainings[16] that are available in most large cities. Your HR department will be able to help you find and select some good courses.

Below is a suggested book to review to help determine the types of program performance indicators you may choose.

> Industrial Hygiene Performance Metrics Manual[17]

Chapter 6

EH&S Management Systems

Every organization's departments should have a management system. This is especially true for EH&S since the department must ensure that it's efficient, effective, utilizes its resources to establish best practices, and keeps the company in compliance with a plethora of various local, state, and federal requirements. Over the last few decades, EH&S management systems have evolved. This chapter recreates some of the evolution to illuminate the functional aspects of a more robust system. One size doesn't fit all. One of the most important tasks of a manager will be to select and put into operation a system that supports their particular corporate structure and business.

Everyone can appreciate that we and our staff need to know, on a consistent basis, what required tasks must be completed. The enlightened manager provides not just a list of requirements, but guidance on how to fulfill requirements. Task management is one of the most important jobs of the EH&S manager. Let's look at some of the ways to develop, continue to improve, and systemize EH&S task management.

Initial Task List

Sometimes the manager is presented with a huge challenge. Let's suppose that he/she is the first full-time manager of an EH&S department. One of the ways that the new manager could determine required tasks is to brainstorm from past experience or meet with a consultant to determine a list of necessary tasks for the team. From this effort, they may create a master to-do list which shows the task, the party to whom the task was assigned, and the necessary completion date. It may also include information used for task ranking, such as:

1. Urgent
2. Important
3. Complete as time allows

Person Responsible	Task to be Completed	Due Date	Completed?
Ben Alt. Jim	Eye Wash Test	6-1-2007	Yes
Larry Alt. Sarah	SCBA Check	6-15-2008	No
Jim Alt. Ben	Lab Inspections	8-5-2008	No
Mike Alt. Jim	CHP Plan Update	12-31-2008	No

The table above is an abbreviated example of the initial list that a manager may make when presented with the challenge of creating a new department from the ground up. This is only a temporary, single-use list. Using this list for an extended period of time would result in problems, because many tasks are recurring. It could, however, be used as a bridge to a more robust system.

In the future, the list may improve by collecting all necessary recurring tasks, identifying the period such as daily, weekly, bi-weekly, monthly, quarterly, semi-annually, or annual tasks, and the primary person responsible to complete the task along with an alternate or backup.

If you put the information together with a calendar, you can map the starting date and the completion period for tasks no matter what the duration of the task.

The type of list below is also called a recurring activities schedule, and it is produced from the initial list. Instead of being temporary, this list is created to be used and updated continuously. Below is an example of an excerpt from an improved task list. Please note the table would include a whole year to the right.

	Period	Task/Activity	Primary	Alternate	January 2001				February 2001			
					1	2	3	4	1	2	3	4
1	D	Unit #1 – Tank inspection	John H.	Carol W.							x	
2	D	Lab waste pick ups	Kevin M.	John H.	x	x	x	x	x	x	x	x
3	W	90-day storage inspection	Carol W.	Kevin M.			x				x	
4	W	Satellite location inspection	Lisa K.	Bob T.	x	x	x	x	x	x	x	x
5	Bi-W	Drum shipment	Bob T.	Larry J.		x		x		x		x
6	M	Eyewash test	Larry J.	Harry G.			x					x
7	M	Scientist safety orientation	Harry G.	Lisa K.			x					
8	Q	Emergency shower test	John H.	Kevin M.								x
9	SA	Lab inspections	Carol W.	Larry J.						x		
10	A	Budget planning	Bob T.	Harry G.								
11	A	Fire drill	Lisa K.	Carol W.								x

Task Management System

A more robust and sustainable system would list the task, your guidelines for completion of the task, and the underlying basis for the task (i.e., the requirement or the actual regulatory citation spelled out or hyperlinked to the task). The explanation should include, in plain language, exactly how to complete the task. The idea is to be able to effectively delegate the task to a staff member, providing them with sufficient guidance to be able to competently complete the task.

Example Task

Task: SCBA Inspection

Frequency: Monthly

Assigned to: Jim Lieberman

Task Description:

Once per month inspect all self contained breathing apparatus (SCBA) using the attached checklist. If the equipment is found to be defective, place an out of service repair tag on the SCBA with your name & date of the inspection. In addition contact Oland Safety to contract for the SCBA repair within 3 days.

Attachment: Inspection checklist

Citation: 29 CFR 1910.134

Inspection

1910.134(h)(3)(i)

The employer shall ensure that respirators are inspected as follows:

1910.134(h)(3)(i)(A)

All respirators used in routine situations shall be inspected before each use and during cleaning;

1910.134(h)(3)(i)(B)

All respirators maintained for use in emergency situations shall be inspected at least monthly and in accordance with the manufacturer's recommendations, and shall be checked for proper function before and after each use; and

1910.134(h)(3)(i)(C)

Emergency escape-only respirators shall be inspected before being carried into the workplace for use.

1910.134(h)(3)(ii)

The employer shall ensure that respirator inspections include the following:

1910.134(h)(3)(ii)(A)

A check of respirator function, tightness of connections, and the condition of the various parts including, but not limited to, the facepiece, head straps, valves, connecting tube, and cartridges, canisters or filters; and

1910.134(h)(3)(ii)(B)

A check of elastomeric parts for pliability and signs of deterioration.

1910.134(h)(3)(iii)

In addition to the requirements of paragraphs (h)(3)(i) and (ii) of this section, self-contained breathing apparatus shall be inspected monthly. Air and oxygen cylinders shall be maintained in a fully charged state and shall be recharged when the pressure falls to 90% of the manufacturer's recommended pressure level. The employer shall determine that the regulator and warning devices function properly.

1910.134(h)(3)(iv)

For respirators maintained for emergency use, the employer shall:

1910.134(h)(3)(iv)(A)

Certify the respirator by documenting the date the inspection was performed, the name (or signature) of the person who made the inspection, the findings, required remedial action, and a serial number or other means of identifying the inspected respirator; and

1910.134(h)(3)(iv)(B)

Provide this information on a tag or label that is attached to the storage compartment for the respirator, is kept with the respirator, or is included in inspection reports stored as paper or electronic files. This information shall be maintained until replaced following a subsequent certification.
1910.134(h)(4)

Repairs. The employer shall ensure that respirators that fail an inspection or are otherwise found to be defective are removed from service, and are discarded or repaired or adjusted in accordance with the following procedures:

1910.134(h)(4)(i)

Repairs or adjustments to respirators are to be made only by persons appropriately trained to perform such operations and shall use only the respirator manufacturer's NIOSH-approved parts designed for the respirator;

1910.134(h)(4)(ii)

Repairs shall be made according to the manufacturer's recommendations and specifications for the type and extent of repairs to be performed; and

1910.134(h)(4)(iii)

Reducing and admission valves, regulators, and alarms shall be adjusted or repaired only by the manufacturer or a technician trained by the manufacturer.

Environmental Management Systems

An environmental management system (EMS) is used to systemize the management of an organization's environmental programs in a comprehensive manner. The EMS may be a standalone system, it may be integrated into an EH&S management system, or even into a more comprehensive corporate compliance system.

An EMS may be modest and resemble an earlier example of a task management system, but for larger organizations it can be complex. It may include the organizational structure, planning, and resources for developing, implementing, and maintaining policy for environmental protection.

An EMS should serve as a tool to improve environmental compliance performance. It should provide a systematic way of managing a company's environmental compliance requirements and affairs, and provide for an orderly and consistent method to address environmental requirements through the allocation of resources and assignment of responsibility.

The EMS Model

EMS is designed as a continuous process, a Plan-Do-Check-Act Cycle. The diagram below shows the process of first developing an environmental policy, planning the EMS, and then implementing it. The model is continuous because an EMS is a process of continual improvement in which a company is constantly reviewing, revising, and tailoring to its needs and goals. As a company matures, so will its EMS.

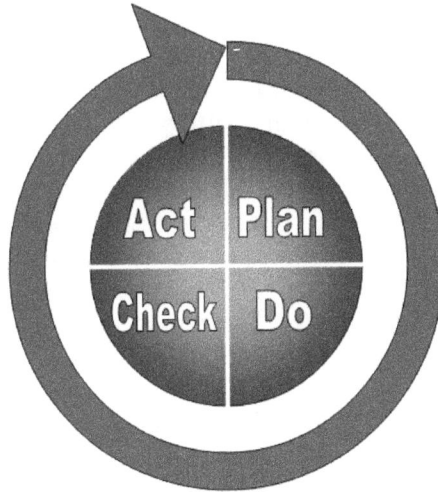

Here are some of the key elements of an EMS:

- A Policy Statement - a statement of the organization's commitment to the environment
- The Identification of Significant Environmental Impacts - from products, activities and services; what are their effects on the environment?
- The development of objectives and targets – what are the environmental goals?

- Implementation – how do we plan to meet objectives and targets?
- Training - instruction to ensure employees are aware of their environmental responsibilities

Guidelines for EMS

The International Organization for Standardization (ISO)[18], located in Geneva, Switzerland, a non-governmental organization, has developed a family of voluntary standards and guidance documents to help organizations address environmental issues. The organization develops voluntary technical standards that aim at making the development, manufacture, and supply of goods and services more efficient, safe and clean.

How Are ISO Standards Developed?

All of the ISO standards are developed through a voluntary, consensus-based approach. ISO has different member countries across the globe. Each member country develops its position on the standards and these positions are then negotiated with other member countries. Draft versions of the standards are sent out for formal written comment, and each country casts its official vote on the drafts at the appropriate stage of the process. Within each country, various types of organizations can and do participate in the process. These organizations include industry, government (federal and state), and other interested parties, like various non-government organizations.

ISO 14000[19] refers to standards and guidance documents to help organizations address environmental issues. Included are standards for Environmental Management Systems, environmental and EMS auditing, environmental labeling, performance evaluation and life-cycle assessment.

In September 1996, the International Organization for Standardization published the first edition of ISO 14001, the Environmental Management Systems standard. ISO 14001 is a specification standard to which an organization may receive certification or registration. ISO 14001 is considered the foundational document of the entire series. A second edition of ISO 14001 was published in 2004, updating the standard.

The U.S. body that provides input for standards development is the Technical Advisory Group (U.S. TAG). This body has established a formal process to respond to questions that may arise regarding clarification of the ISO 14001. Responses will reflect the interpretation of the Standard as intended during the drafting of the Standard, and may be found in the "Clarification of Intent of ISO 14001."

An EMS should be flexible, and does not necessarily require a company to change its existing activities. In actuality, however, if the use of the system has no impact on the company's programs it is a futile effort. A well developed EMS establishes a management framework by which an organization's impacts on the environment can be systematically identified and reduced. Although the standard was not exactly designed to be used to assist with maintaining compliance, it is an invaluable tool for doing just that.

The following are the specific requirements of ISO 14001:

- Environmental Policy - develop a statement of the organization's commitment to the environment
- Environmental Aspects and Impacts - identify environmental attributes of products, activities, and services, and their effects on the environment
- Legal and Other Requirements - identify and ensure access to relevant laws and regulations
- Objectives and Targets and Environmental Management Program - set environmental goals for the organization and plan actions to achieve objectives and targets
- Structure and Responsibility - establish roles and responsibilities within the organization
- Training, Awareness and Competence - ensure that employees are aware and capable of their environmental responsibilities
- Communication - develop processes for internal and external communication on environmental management issues
- EMS Documentation - maintain information about the EMS and related documents
- Document Control - ensure effective management of procedures and other documents
- Operational Control - identify, plan, and manage the organization's operations and activities in line with stated policy, objectives, and targets

- Emergency Preparedness and Response - develop procedures for preventing and responding to potential emergencies
- Monitoring and Measuring - monitor key activities and track performance including periodic compliance evaluation
- Evaluation of Compliance - develop procedure to periodically evaluate compliance with legal and other requirements
- Nonconformance and Corrective and Preventive Action - identify and correct problems and prevent recurrences
- Records - keep adequate records of EMS performance
- EMS Audit - periodically verify that the EMS is effective and achieving objectives and targets
- Management Review - review the EMS

Legislation and Standards

If you do business in the European Union, the Environmental Liability Directive [ELD] 2004/35/EC is one of the most important instruments that your business will need to comply with, and must be included in your EMS. In March of 2009 it came into force in Europe, superseding the various national pollution prevention guidelines. Failure to comply can result in fines.

Within this Directive is a requirement to mitigate the effects of such events as spills and firewater (runoff from fires). The Directive makes it clear that it is the site owner's responsibility to contain spills and firewater on site using some form of containment apparatus such as sealing the drains.

The implementation of a robust EMS will incorporate ISO 14001 requirements and should lead to improved environmental performance, including better and more consistent legal compliance.

A Well Crafted EMS

- Serves as a tool to improve environmental performance
- Provides a systematic way of managing a corporation's environmental affairs
- Is the aspect of the organization's overall management structure that addresses immediate and long-term impacts of its products, services, and processes on the environment
- Gives order and consistency for organizations to address environmental concerns through the allocation of resources,

assignment of responsibility, and ongoing evaluation of practices, procedures, and processes
- Focuses on continual improvement of the system

Mature System

A mature system for tracking EH&S tasks should be an electronic database with reporting capability that allows a manager to schedule a task, assign personnel to perform the task, identify the performance period, and track and document completion status. If you have recurring tasks, they should be automatically regenerated according to the required frequency needed.

An excellent example of a mature system for the life sciences industry is Affytrac. Below is what the vendor has to say about the software.

Affytrac is our easy-to-use, affordable, secure EH&S compliance management system for the life science industry. As a web-based tool, there's no software to install, and with our streamlined setup process, there are no delays in having your EH&S program automated, optimally tuned, and always at your fingertips. All you need is a web browser. Simple and intuitive to use, yet powerful and full-featured, **Affytrac** includes all the functionality you expect in an EH&S management system, without any unnecessary complexity to get in your way.

- Task Management lets you define your organization's set of compliance and safety tasks. Assign to personnel, add your target dates, and Affytrac automatically notifies staff and tracks progress.
- Document and track Corrective Actions identified through accidents, incidents, safety committees, regulatory inspections, and risk assessments.
- The Potent Compounds repository tracks the stage of development, therapeutic indication, facilities where the material is used, potent compound classification, and more. Attach potent compound classification reports, material safety data sheets, and industrial hygiene reports.

- Our large library of over 200 **fully-documented compliance task templates** lets you tailor and put your program in motion on Day One.

- Our Potent Compound Safety Self-Assessment Tool (PCSSAT) lets you evaluate your potent compound safety program against best-in-class criteria. Learn from the experts without the need for time consuming and expensive assessments performed by third parties.

- Quickly produce built-in reports and graphs, or export data to spreadsheets for more complex analysis.

Additional References:

The following books are good references to help you establish an EH&S management system:

OHSAS 18001: Designing and Implementing an Effective Health and Safety Management System[20]

ANSI/AIHA Z10-2005 Occupational Health and Safety Management Systems[21]

The following organization may provide the member with significant assistance with management systems:

The National Association of EHS Management (NAEM)[22]

Chapter 7

Introduction to Basic Business Concepts

This chapter will help you to understand and begin to use the language of business within your organization. It will offer suggestions for ways to collect financial information you will need to complete annual budgets and support requests. To objectively present and defend your department's financial requirements, you need data. The next chapter will assist you in turning the financial data you have collected into business proposals for obtaining financial resources.

Management of Resources

It bears repeating that profitability is indispensable to business success. A lack of profitability shows poor management. Therefore, profit is the normal basis for business decisions and the customary gauge of business performance.

So how does this affect the manager of a department such as environmental health and safety? Generally, it puts him/her at a disadvantage. Normally administrative departments do not generate a profit; they are almost always an expense. It will always be difficult to get management's attention and appreciation when your department can never be a profit center.

But before you become resigned to this unhappy state of affairs, let's take another look and see what opportunities present themselves. First and foremost, the manager must appreciate the fact that his department, since it cannot generate funds, can only positively impact the company's profitability by avoiding ineffective expenses.

We need to look at short term and long term expenses. Your group is a value-added department if it gets the job done and avoids unnecessary expenses. Put another way, once we have an operating department, we're looking to get the highest marginal benefit for the company for the least marginal cost.

Let's look at the analogy in a few different ways. If we look at a 1950s Ozzie and Harriet image, Harriet stayed home, shopped, and took care of Ozzie and the kids. She did not have an opportunity to earn money, but by intelligently shopping she could avoid some costs.

Let's look at an example where she can spend more on a specific purchase, yet save money for the family in the long run. Let's say a pair of regular blue jeans for the kids costs $15.00 and a pair of blue jeans with reinforced knees costs $17.00. What would be the better value? Well, Harriet knows from experience that the regular jeans would last 12 months. Her friends have told her that the jeans with the reinforced knees would last 6 months longer, or 18 months. Which is the better value?

Regular	$15/12 = $1.25/month
Reinforced	$17/18 = $ .94/month

Harriet, being the good shopper, spends the extra $2.00 but gets a better value. We know that Ozzie already loves her, but he will better appreciate her efforts if she explains the $0.31/month savings to him – especially if she annualized the savings ($0.31 x 12 = $3.72/year).

Now let's substitute an administrative department, in this case EH&S, for Harriet, and Management for Ozzie. Next substitute day-use canal caps for reinforced blue jeans and single-use disposable hearing protection for regular blue jeans.

To make the most of the Ozzie and Harriet analogy, we must first recognize the assumptions. That is, we need to have good information concerning the life of the blue jeans. What would good information look like? Well if Harriet asked her three other bridge club members how long the jeans lasted, she would have a better justification for the life of the jeans. We will expand on this theme in the next chapter.

Using Financial Data

To create the graphs that turn data into information, you need to collect the data. In larger, more established companies this will have been set up for you, but in smaller companies it may not be well established.

Since EH&S expenses are typically less than 1% of total corporate expenses, many accounting departments do not or cannot provide the detail necessary to make intelligent day-to-day decisions. This puts the department manager in a tough position. You don't get to recreate the accounting system, but you do need better data. What is the answer? I suggest that you do the following:

1. Keep a copy of all contracts and agreements you negotiate in your files.
2. Save all of the invoices that your department approves in your file.
3. Create a budgeting spreadsheet.

Save the copies of the contracts and agreements in an electronic file. Scan documents as necessary.

Make sure that you save the signed copy. A good system for organizing is to save the files under year, then under name or functional area: e.g.,

2009
    Clean harbors
        Hazardous waste disposal

A similar system can be created and used for filing invoices. Some companies like Array have a different fiscal year than calendar year. You may need to take this into consideration when creating your filing system.

If you save the agreements and invoices for at least one year, you will be in a position to create a budget expense spreadsheet. I have my budget spreadsheet organized by functional area, listing vendors under the headings. Then I show last year's budgeted amounts. Next I have space to fill in twelve months of expenses and, at the end of the row, the months are summed. Refer to the budgeting expense example in chapter 9, page 72.

The lesson for the EH&S manager is to collect good information on price and product (service/project) life. Product life is a much more difficult value to obtain, but you can get if from:

1) Historical records
2) Available publications
3) Peer group
4) Industry group
5) Vendors

Obviously the time/cost outlay to obtain good information should be proportional to the magnitude of the transaction.

Rule of Thumb: You should collect enough data from reliable sources for you to comfortably justify the values.

Department Expenses and Year to Year Changes

It is important for the EH&S manager to track his department's expenses. Obviously we need to know the actual sum per year, but the changes in costs are very important. One useful tool to use is percentages. Let's look at a few examples.

If our hazardous waste disposal costs were $100,000 in 2008 and $110,000 in 2009, the percentage increase in cost would be equal to: 2009 expense, minus 2008 expense, divided by the 2008 expense, multiplied by 100, equals percent change.

$$\frac{(\$110,000 - \$100,000)}{\$100,000} \times 100 = 10\% \text{ increase}$$

It is important to realize that costs are increasing, but if we do not normalize the gross sums we may be missing very important information. How do we normalize? We may do it against production. So if we made 100,000 widgets in 2008 and 110,000 widgets in 2009 then:

$\frac{\$100,000}{\$100,000}$ = $1.00 of hazardous waste expense per widget for 2008

$\frac{\$110,000}{\$110,000}$ = $1.00 of hazardous waste expense per widget in 2009

So in reviewing the example, the hazardous waste expense did indeed increase in gross terms, but by unit of output, it did not. It is important to be able to identify dependent vs. independent cost. The usual terms used for dependent cost are variable costs, costs that vary depending on production or other business operations, whereas fixed costs, such as rent, are independent of production or business operations.

Items, for which we should track gross cost, year to year changes, and variable cost, may include the following:

Chemical waste disposal
Contracted services

Emergency monitoring
Equipment calibration & repair
Ergonomic equipment / evaluation cost
Fire protection systems, inspection / repair
First aid supplies
Lab sample analysis
Medical monitoring
PPE
Professional fees
Training
Travel
Workers' compensation insurance

Managers may also do other trending analysis. For example, one may want to see waste disposal cost per R&D dollar spent as a percentage or as a dollar per dollar expense.

It most likely would be difficult or impossible to track individual program cost. We can track the EH&S department's yearly costs and we could normalize by the number of company employees, the number of scientists in an R&D environment, or the number of production workers in a manufacturing environment. This requires the EH&S manager to work with accounting to get relevant annual expenses.

The types and ranges of accounting data collected will vary tremendously depending on the size of the company, type of industry, stage of maturity, and corporate accounting systems. One of the major responsibilities of an EH&S manager is to review his company and department, objectively identify corporate goals, and how his department goals and objectives will support the corporate goals. He needs to convert the department's goals and objectives into discreet identifiable doable tasks. He should be able to estimate the cost for newly identified tasks prior to their initiation.

EH&S Managers should follow this rule: Never collect data if you will not use it or if you have no standard of measurement for it. The Les McCann song has it right on: *Compared To What?* If you collect data make sure that you have some comparison scale to use to interpret and turn the data into valuable information that empowers you and your corporate management to make better decisions.

Chapter 8

Financial Tools

Next, we turn to the business financial tools you'll need to turn the data you have collected, or will collect in the future, into compelling financial arguments. If you don't have a financial educational background, that won't be an insurmountable handicap. I hope that the descriptions, explanations, and examples will allow you to grasp the concepts and use them to your advantage. Remember, you need not work in a vacuum. If you get insight from this chapter, you may ask to work with someone in your accounting department to flesh out your arguments. But you will be on your way to using and presenting financial data to support your department's needs and programs.

When reporting to executive management you must make your reports include returns on investments expressed in terms of money. Capital is the language of business.

Below is a list of essential concepts that we will review:

- Capital/Resources
- Capital Budgeting
- Gauge of business performance: Return on Investment
- Time Value of Money
- Cost of Capital
- Ranking and selecting capital projects

Capital Budgeting

Most EH&S managers need to annually budget for their department. The exercise usually requires cost estimations for necessary services and requests for additional funds for new equipment (capital budgeting) or service maintenance. Almost all management scrutiny, evaluation, and required justification from the EH&S manager involves expenditures for new equipment, services, or programs. My discussion will focus on the justification for new equipment services or programs.

Return on Investment

If one was to invest $1,000 at a simple interest rate of 10% per year, then the return would be:

$$\$1,000 \times 0.10 = \$100 \text{ per year}$$

If your company invested $500,000 and made a profit of $100,000 for the year, assuming that the initial investment will be repaid, then the return on investment would be:

$$(\text{Profit on Investment / Dollars Invested}) \times 100 = \text{Return in Percent}$$
$$(\$100,000 / \$500,000) \times 100 = 20\%$$

Typically the EH&S manager will not be presented with the above type of investments, rather the return on investment will be a savings.

So let's reframe the example. If we spend $10,000 per year on disposable PPE lab coats and we found a laundry service that cost $5,000 per year, but requires a one-time purchase of lab coats of $15,000 should we make the investment? We need to know the length of savings. In this case the agreement with the contractor is for a period of five years. Let's start with calculating a payback period.

Payback Period

Savings per year = (Current Expense - Future Expense)

Savings per year ($10,000 − 5,000) = $5,000

Initial Investment / Yearly Savings = Yrs

$$\$15,000 \ / \ \$5,000 \text{ per year} = 3 \text{ years}$$

Rate of Return

So the prospect looks reasonable. How would we calculate a specific rate of return for the project?

$$\$5,000 \text{ per year x } 5 \text{ years} = \$25,000 \text{ total savings}$$

(Total Saving - Initial Investment) / (Initial Investment) x 100 = Return in Percent

$$(\$25,000 - \$15,000) \ / \ \$15,000 \text{ x } 100 = 66\% \text{ Return on Investment}$$

Usually we are interested in an annualized return, so the simple interest or annual rate of return would be:

(Yearly Saving - Yearly Investment) / Initial Investment x 100 = Return in Percent

$$(\$5,000 - (15,000/5)) \ / \ \$15,000 \text{ x } 100 = 13\%$$

Net Present Value

For a project that we have described we could also take into consideration the time value of money and calculate the net present value (NPV) of the investment. But to do so, we would need the cost of capital. For example, let's assume the company can borrow money for the five year period at rate of 8% per year. So including the time value of money, the project needs to make at least 8% per year to break even. For this project we will call the 8% the Project Required Rate of Return.

The formula for the NPV is:

$$NPV = \sum_{t=1}^{n} \text{x} \frac{CF_T}{(1+R)^t} - I_o$$

Where
NPV   = net present value
CF_T = sum of the annual cash flow (Savings)
R   = required rate of return (Cost of Money)
I_o = net money invested (Initial Investment)
t   = years
n   = number of years

Let's show the calculation:

$$NPV = \frac{5,000}{(1+.08)^1} + \frac{5,000}{(1+.08)^2} + \frac{5,000}{(1+.08)^3} + \frac{5,000}{(1+.08)^4} + \frac{5,000}{(1+.08)^5} - 15,000$$

$$NPV = 4629 + 4287 + 3969 + 3676 + 3403 - 15,000$$

$$NPV = 19,964 - 15,000$$

$$NPV = \$4,964$$

So, another way to think of the NPV is to say that the investment will produce a net return of $4,964 considering the cost of money.

Internal Rate of Return

Let's use the same example, but this time we will calculate the investment's rate of return. The internal rate of return (IRR) is the discount rate that equals the present value of a projects total cash inflow minus its net investment cash outflows. Another way to say it is the rate that makes the projects NPV = zero.

Formula is:

$$I_0 = \sum_{t=1}^{n} x \ \frac{CF_t}{(1+IRR)^t}$$

So:

$$15,000 = \frac{5,000}{(1+IRR)^1} + \frac{5,000}{(1+IRR)^2} + \frac{5,000}{(1+IRR)^3} + \frac{5,000}{(1+IRR)^4} + \frac{5,000}{(1+IRR)^5}$$

This calculation is best done with Excel or a financial calculator, but it could be done by trial and error. We know that the rate must be greater than 8% by our previous example; by using an Excel spreadsheet and plugging for the IRR, the value is found to be approximately 0.198 or 20%.

Another way to look at the IRR is that not only will the project pay for the money borrowed to fund it; it will provide a rate of return of 12%. Many companies have a minimum IRR for projects to be approved or they may use the IRR to rank projects.

I realize that in practice your particular investment decision may be more complicated, but if you take the time and effort to collect the necessary information and calculate a payback period, rate of return, net present value, and the IRR, you will significantly improve your ability to review projects, improve the prospects for projects to be approved, and be looked on as a fully participating member of management.

Cost of Insurance

Normally we purchase insurance to provide protection in the case of unexpected events. We can use this model to our advantage to compare the cost of prevention of unwanted events. Most insurance has a deductable amount, (out of pocket, and sometimes a copayment). There are also exclusions.

Insurance costs vary dramatically depending on the item or intangible to be insured. Prices for insurance range from a fraction of a percent of the value of the item or intangible up to about five percent. After five percent, you are no longer talking about insurance, but placing a bet.

Let's look at an example. In this case, term life insurance. Let's assume you are a forty year old male in good health and a non-smoker. To purchase a five hundred thousand dollar term life insurance policy, you would need to pay approximately thirty-five dollars per month or about four hundred and twenty dollars per year. Therefore, the percentage of the intangible (your life) that you are insuring per year would cost.

$$(\text{Annual Insurance Cost} / \text{Insured Amount}) \times 100 =$$
$$(\$420 / 500{,}000) \times 100 = 0.084\%$$

Now let's look at the odds of you dying in the next year. According to the US Social Security Administration actuarial life table, your death probability is 0.2391% during your fortieth year. So let's compare: it looks like you're paying 0.084% to cover a risk of 0.2391%. I think that most rational people would agree the cost of life insurance is worth the protection it affords. So, how can the insurance company offer a percentage rate that is below the percentage possibility of your death? If you think about it, the insurance company sets exclusions based on your current health status. These exclusions must reduce the chances of your death in the next year from 0.2391% to below 0.084%.

In fact, I believe that the real risk the insurance company plans for is less than 0.009%. They're in the business of making money, not losing it. But before we get off track, the insurance company does what good

EH&S managers do, reduces risk by implementing administrative and engineering controls.

In this case, they're using administrative controls to limit the pool of insured males who have a much greater likelihood of dying within a year than the normal forty year old male. Before we leave this example, the take home message is that it is reasonable to pay approximately 0.08% to protect the value of an intangible.

Let's look at another insurance model example. This time, homeowner's insurance. Let's assume that your home is worth $305,000. The insurance cost per year for the home would cost approximately $1,000 per year with a $5,000 deductable. So the percentage cost would be:

$$(1,000 / (305,000 - 5,000)) \times 100 = 0.33\%$$

Most reasonable people would agree that paying 0.33% to insure the value of an asset as necessary as your home is a reasonable investment.

Let's apply these insurance examples by analogy to an EH&S department. You believe that your facility is in compliance with federal, state, and local regulations. To assure that your facility is in compliance, you would like to contract for an EH&S audit. The audit would cost $10,000. You estimate that the aggregate non-compliance fines could cost up to $250,000 per year for your facility. Should you contract for the audit?

$$\text{(Cost of audit / potential aggregate fines)} \times 100 =$$
$$(10,000 / 250,000) \times 100 = 4\%$$

The percentage cost of the audit is four percent of the cost of the potential aggregate fine. Therefore, the cost of insuring against a loss (fine) in this case is four percent. A four percent cost of insurance is on the top end of the reasonable cost of insurance. Therefore, this particular decision should be evaluated carefully.

Let's add an additional factor. Instead of an annual audit, you may plan for a biennial audit, one every two years. This would reduce the cost by fifty percent per year. Theoretically, the value of the first audit should be greater than those following, since you would expect most deficiencies to be identified and corrected within the first year. Considering a biennial audit may be a reasonable compromise while still providing the review needed. This is a question that would be perfectly suited to be examined by your superior or executive management to determine their risk avoidance and tolerance levels for the organization. However you, the EH&S manager, have provided the financial framework to pose the appropriate question for consideration and evaluation.

Ranking and Selecting Capital Projects

Where applicable, the payback period method should always be applied by the EH&S manager. The simple payback period calculates the time required to recapture the initial investment from a project's net investment cash inflows. For example, if a company would like to invest $120,000 in the first year on a bulk hazardous waste storage facility that saved $20,000 per year in disposal cost, what would be the payback period?

$$\$120,000/\$20,000 \text{ per year} = 6 \text{ years}$$

Usually projects with a payback period of greater than 3 to 5 years are rejected. This is where the EH&S manager must be sure he has captured all the savings. For example, perhaps the additional freed up space in the building can be put to other uses. Perhaps the more secure storage would reduce the annual insurance costs. These additional savings should be estimated and included.

So if:

Freed up space = 500 sq.ft at $15 sq.ft per year = $7,500/year

Reduced pollution liability insurance = $2,000/year

We now have a direct savings of $20,000/ year, plus the freed up space savings of $7,500, plus reduced insurance costs of $2,000/year for a total savings of $29,500 per year. Now we revisit the payback period.

$$\$120,000 / \$29,500 \text{ per year} = 4 \text{ years}$$

Now the full savings reflect a payback period of 4 years. The project may still be rejected, but it is more reasonable. This is the point at which more intangibles may be mentioned, such as a decrease in hazardous material storage risk or the possibly of less onerous state and federal reporting requirements. You could always estimate dollar values for savings that the project may realize using the most likely case.

Economic Life vs. Operational Life

It is very important to select an appropriate life span for an investment. I would suggest that the range should be from one to ten years, with the majority in the three to five year period. Let's explore an example to see why. The economic life is the period of time during which an asset competitively produces a good or service of value. The economic life of an asset may be particularly short in a rapidly changing field such as electronics where new developments often render an asset obsolete shortly after it is purchased.

Example Economic Life for PC Purchase

If we purchase a PC today for $2,000 that PC may physically operate for five years or possibly more, but with the evolution of software, hardware, and systems capability, the computer will most likely reach its useful economic life in 3 years. If we were going to do an NPV calculation we should use a maximum life of 3 years (a shorter life may even be appropriate for quickly evolving technology areas).

Your company will have guidance on the depreciation period for assets. Ask the accounting department for a copy of the guidance. The economic life should never exceed the assets' depreciation period.

Many of my fellow EH&S managers work in an early or middle stage startup; companies that have yet to turn a profit. These R&D companies must use their debt and equality funds very efficiently, but also in a very entrepreneurial way. So the conservation of capital and flexibility may many times outweigh future savings. In that environment, usually management looks for a three year or less payback period.

Presenting a proposal to executive management must be in terms of dollars and cents. It isn't always easy to put your idea in terms of capital.

Example of Valuation for Vaccinations

A manager is deciding whether they should or shouldn't recommend that their company provide free flu vaccinations to employees.

In order to answer this question, you have to collect costs and make your assumptions. Any time you have to make a business decision, you need to come up with the data and assumptions. In this case, you would start off by finding the cost of vaccination and how many people are in your company.

Cost of vaccination=$25   Number of Employees: 350

After you have established the basics, you should determine the time period a person would be absent from work if they catch the flu. The Internet is a great resource to be used to your advantage to find answers. Appropriate searches will yield at least some salient information to almost every question, such as the one being posed in the previous sentence. Within minutes of searching for this information, the answer presented itself.

Average time absent from work: 3 days – 24 working hours[23]

The next bit of information you need is the per hour, fully loaded average employment cost for personnel, specific to your company. This can be achieved by asking the accounting department of your company.

Personnel average per hour rate: $35 per hour

The next thing that must be factored in is the time employees would be absent from work in order to receive their vaccination. For this piece, a rough estimation will suffice. Make sure you factor in travel and other important aspects.

Time to take vaccinations: 1 hour

More information from the Internet is necessary to calculate the cost of the vaccinations. You should find the odds of getting the flu,[24] and make sure the age range is correct for your company's employees. For example, the statistics I found were:

Odds of getting the flu: 5% - 20%   Age 18 – 49: 7%

Because most of the company's workforce will be between ages 18 and 49, much more accurate data can be collected on the odds of getting the flu. The next thing to find out is the rate at which flu vaccinations reduce the risk of a healthy adult getting the flu. This is another fact to find on the Internet. If you get a range of numbers and you can't narrow it down specifically to your company anymore, it's best to find the average of the two numbers.

Flu vaccinations reduce the risk of a healthy adult getting the flu by a rate of 70% - 90%, so the average is 80%.

The last thing that you need to factor into your idea is the percent of your company's employees that would actually take advantage of the free flu vaccinations. This number should be determined by your own personal experience, or that of your professional network.

Percent of employees actually getting flu shot: 50%

After you have collected the essential information, you can start your calculations. The first thing to calculate is the cost of delivery of the vaccinations. To do this, the following equation is used:

(Cost of each vaccination   x   # employees) + ((Time to take vaccinations x # employees) x employee cost per hour x % employees vaccinated)

Substitute the numbers you found:

($25 x 175 employees) + (1 x 175 employees x $35/hr) = $10,500

The next piece to calculate is the loss to the company if no vaccinations were to take place:

(# employees x avg. rate per hour x time missed from work) x % of adults who catch the flu

(350 employees x $35/hr x 24 hours) x .07 = $20,580

After finding these two numbers, you should add back for the percent of vaccinations that are ineffective. This is another piece of information where personal experience and the Internet will yield an answer.

20% of typical flu vaccinations are ineffective.

Now, calculate the loss to your company because of the ineffective vaccinations:

(# employees x average per hour rate x time absent from work) x (percent taking vaccination) (% ineffective) x (% odds of getting flu)

(175 employees x $35/hr x 24 hours) x (.2) x (.07) = $2,058

We still have to account for the 175 who were not vaccinated:
(175 employees x $35/hr x 24 hours) x   (.07) = 10,290

After calculating all the above numbers, the total savings can finally be determined:

> Loss to the company if no vaccinations are given
> − (Cost of vaccinations + cost from ineffective vaccinations+ loss from those not vaccinated)

$20,580 − ($10,500 + $2,048 +10,290) = total loss of -$2,268

Executive management will now have financial basis to accept your idea, especially if the payback period is short and financial analysis shows a positive return.

Payback Period

Let's calculate the payback period for the above example:

Initial Investment = (Cost of vaccinations) =$10,500

Initial Investment / Total Savings = Yrs

$10,500 / $-2,268 = Loss: so payback not applicable

Let's assume that you could reduce the time for vaccinations to be ½ hour. Then:

Cost of vaccination:

($25 x 175 employees) + (175 employees x $35/hr x 0.5) = $7,438

With ½ hour time:

$20,580 – ($7,438 + $2,048 +10,290) = Total Savings of $804

Payback Period $7,438 / $804 = 9.25 Years

Rate of Return

So the prospect doesn't look overly appealing. How could we improve the chances of this being accepted?

Possibilities:

1) Administer vaccinations before or after work hours and have no loss work time? If so then:

$20,580 – ($4,375 + $2,048 +10,290) = Total Savings of $3,867

Now you are armed with the financial information needed to make your argument to management. The vast majority of executive managers will approve a proposed investment that has a less than a three year payback period and a positive rate of return.

Additional References:

You may find the following books to be helpful resources:

Finance for Non-Financial Managers[25]

Finance for Managers[26]

Chapter 9

Budgeting

Now we turn our attention to annual budgeting. This is a task few managers relish, but all are required to complete. If you have followed the advice in chapters seven and eight, you will be armed with the information to put together a reasonably prospective annual budget. Budgeting comes in two flavors: operational and capital. Operational budgets lay out the expected expenses you foresee to keep current programs funded, and capital budgets are designed to force the manager to specifically address requests for new and substantial funds. You'll be expected to make a compelling case for a new program or request for a capital item. The tools and examples that you have seen in this book should assist you in making your argument.

Operating Budget

All well run businesses operate on a budget. You will be responsible for creating or contributing to your annual budget. The annual budget incorporates your department's expected expenses for the next year. For larger projects you would include a capital budget. A budget is a forward projection and an expression of your business. It serves to quantify the expected expenses and delineates estimates for your department's conceptual ideas, transforming them into reviewable costs.

A well-prepared budget will serve as a tool to communicate with your company's accounting department and executive management. It will establish the baseline and a reference point to guide your staff by providing an essential framework for what purchases your department expects and has planned for during the next year. It is a very important tool in managing your department, provided it is based on realistic data that can be used to gauge actual performance.

Your department's budget most likely will be one component of a larger detailed corporate wide budget that your accounting department will create and executive management will review. The corporate wide budget will be used to determine capital needs so that executive management can present accurate projections for lenders and investors. Be proactive; take the accounting department's manager to lunch. Ask how accounting is tracking your department's expenses, offer suggestions. Then go and think things through. Will your company's accounting department be able to deliver the expense information you need at the level of detail that is required? You will need some time and experience to answer that question. My experience is that most accounting departments want to provide the level of detail that you may want, but they need to appropriately code for the expenses. It will be up to you to present an example of what you seek. Examples presented in this book will help you to convey what information you would like to gather. Be patient; it will probably take some concerted effort by you and the accounting department to get it right. Remember: the more information that you can track, the better able you will be to identify trends.

Building Blocks

A budget is an estimated projection of future expenses. It should be based as much as possible on actual data obtained by looking at past expenditures, doing research, getting quotes, and making estimates of future costs for new projects. A budget should be built from the bottom up. The idea is to start with what you know or what you can find out. For the EH&S manager, starting at the bottom means assembling basic expense details for the expenses you can foresee for the next year. You will need to assemble the basic building blocks to construct the various different line items, such as payments, to service contractors, suppliers, equipment rental, and other overhead expenses. For the typical EH&S department the following expenses are expected: waste disposal, office supplies, safety supplies, medical costs, worker's compensation, industrial hygiene services, licenses and permits, professional fees, conference costs, training, reference materials, and meals and entertainment.

The building blocks are quantities, hours, prices, and rates. It is best to try to match amounts and the time period when the expense is expected to be paid. So if you budget a certain cost for a service by the month the total cost will be the sum of the costs per month.

If a department's business involves jobs or projects, budgeting will probably include aspects of both capital and service budgeting. Based on the particular nature of your business, you will know what is involved in completing jobs or projects. The work may involve materials and supplies, direct and indirect labor, use of equipment, and possibly subcontractors. Budgeting for jobs or projects may go hand-in-hand with submitting job estimates or bids. The terms and conditions of the work will play an important role in how the work is quoted or budgeted. A lump sum contract for a particular job is different from a fixed rate contract.

Breaking down the job into its component parts will be a key factor in submitting a good quote or bid, and budgeting job costs and revenues. You will want to make sure you have all your bases covered in the sense that you include all your costs, both direct and indirect. You may need to make out a materials list, a job schedule for direct labor, and an allocation calculation to include indirect expenses.

If you are going to use subcontractors you should get bids from them, hopefully before committing yourself to the job.

Employee compensation can be budgeted based on the hourly wages or monthly or annual salary contracted with each person. Most likely this cost will be handled by your accounting department, but if not remember your budget will need to include employer payroll taxes, which can be calculated using the applicable rates for Social Security, Medicare and unemployment taxes. You will also need to provide for any benefit plans you have agreed upon with your employees, such as health insurance, pension plans, and others. If you have actual data on the cost of these plans, all the better. If not, you may want to budget based on a payroll burden, which is a percentage of employee based compensation.

You may have other employee-related expenses such as bonuses and awards, which are tied to the completion of objectives or goals. It may be possible to correlate these expenses with a related item in the budget, such as revenue.

You probably know what types of insurance coverage your business will need, and you may be able to get quotes on-line. Or you could go to an insurance agent and inquire. Insurance is generally prepaid, and you can do some accounting to spread the quoted insurance premium month-by-month over the policy period. If you have a periodic preventive maintenance contract on equipment, you can budget based on the monthly cost of your contract. If you have purchased an extended warranty, you can spread the cost over the expected useful life of the related equipment, or the duration of the contract. Repairs are generally an unforeseen expense and are more difficult to predict and budget, but they should not be overlooked. You should probably estimate repair expenses based on your own knowledge and experience, and based on the age and general working condition of your equipment and facilities. If you or your department members use their own vehicle in the business, you can estimate the average number of miles or kilometers you expect to drive each day or month. Then take this and multiply by the appropriate factor based on what the IRS repayment rate is for the type of vehicle. To calculate travel expenses, you will want to collect information by estimating how many trips you plan to take to what places, and find out the going airfares and hotel rates.

As mentioned above, these are just some examples and ideas, and you will need to adapt them to your own particular circumstances. The important thing is to analyze your expenses, and break them down into pieces of data that can be identified. Budgets are not all the same, and there is no standard format. Different types of budgets can be prepared for different purposes. One way to start is to prepare two separate budgets – one for operating costs and one for capital expenditures.

Below is an example of a spreadsheet for an operating budget.

				FY 2007				FY 2008
					Jan	Feb		
1	Hazardous Waste Disposal							**$140,000**
	a	HW Vendor1		$ 72,000	8,000	10,000		$72,000
	b	HW Vendor2		$ 40,000	4,000	0		$40,000
	c	HW Vendor3		$ 26,000	2,167	2,167		$28,000
2	Suppliers							**$27,000**
	a	ABC Chemicals		$ 12,235	1,080	1,080		$12,960
	b	Drun Sales		$ 2,214	185	185		$2,214
	c	ABC First Aid		$ 3,150	300	300		$3,600
	d	GG First Aid		$ 300	25	25		$300
	e	Ergo Products		$ 1,920	160	160		$1,920
	f	Safety Shoes						
		i	Wing	$ 1,000	82	82		$984
		ii	High	$ 350	35	35		$420
	g	Opto (Safety glasses)		$ 600	50	50		$600
	h	B Plus		$ 1,200	100	100		$1,200
	i	Rad Badge Monitoring		$ 3,000				$2,802
3	Medical Cost							**$30,000**
	a	Tree		$ 19,000	1,600	1,600		$19,200
	b	WW		$ 2,000	167	167		$2,000
	c	NC		$ 1,200	100	100		$1,200
	d	ART		$ 1,200	100	100		$1,200
	e	NJ		$ 5,500	500	500		$5,130
	f	ProMed		$ 1,250	500	70		$1,270
4	BioHazardous Waste							**$19,000**
	a	Scycle		$ 17,600	1,563	1,563		$18,790
5	Radioactive Waste							**$8,000**
	a	RWV		$ 8,000		8,000		$8,000
6	WorkersÕCompensation							**$86,000**
	a	Ton		$ 6,000	500	500		$6,000
	b	Pco		$ 80,000	6,667	6,667		$80,000

7	**Air Monitoring**					**$6,500**
	a	MMA Consulting	$ 5,250			$5,250
	b	BV	$ 400	100	100	$1,250
8	**BioHoods**					**$6,600**
	a	HSS	$ 3,600	300	300	$3,600
	b	TSS	$ 3,000	250	250	$3,000
9	**Services**					**$115,000**
	a	Fire Monitoring & Maintenanc F-B	$ 10,560			$10,560
		ii F-L	$ 4,590		1,100	$4,600
		iii F - Service	$ 21,000	500	500	$6,000
	b	After hour emergency calls				
		i Page	$ 720	60	60	$720
	c	Laundry				
		i G&K				
		1	$ 21,000	2,383	2,383	$28,600
		2	$ 44,000	3,423	3,423	$41,080
	d	Equipment Calibration				
		i Heck-Tech	$ 390	200		$400
		ii Shortridge	$ 700			$700
		iii TSI	$ 200			$200
		iv Thermo	$ 2,500	2,500		$2,500
		v Quest	$ 150			$200
		vi Oilind	$ 400			$400
		vii Ludlum	$ 1,080			$1,200
	e	Temp. Employee				
		i Temp	$ 1,642			$1,642
	f	Ergonomics				
		i CC	$ 1,500			$1,500
	g	Recycling Services				
		i Eco-Cycle	$ 2,500	200	200	$2,400
		ii MT	$ 1,000		1,000	$1,000
	h	Consulting Services &				
		i H & A	$ 1,200	100	100	$1,200
		ii AF	$ 4,000			$4,649
		iii CPR Pro	$ 3,000	250	250	$3,000
		iv Applied Rad	$ 1,000			$1,000
	i	AED servicing				
		i StatP	$ 125			$125
	j	Transportation Spill				
		i Chemtrec	$ 675			$1,324

10	Licenses and Permits							$14,000
	a	CDPH&E						
		i	Hazardous Waste	1	$	3,410		$3,500
				2	$	3,410		$3,500
		i	Rad M		$	4,405		$4,625
		iii	Rad E		$	1,000		$1,000
	b	Dept of the Interior			$	200		$200
	c	DOT			$	775		$775
	d	DEA			$	400	400	$400
11	Professional Associations							$2,450
	a	CSA			$	500		$735
	b	BAC			$	1,600		$1,600
	c	ABIH			$	115		$115
12	Training / Conferences							$3,300
	a	AIHce			$	1,500	1,500	$1,500
	b	EHS BP			$	1,800		$1,800
13	Reference Materials				$	2,400	200	200	$3,600
14	Meals and Entertainment				$	2,400	200	200	$2,400
						F08 Total Budget		$463,850

Capital Budgeting

If you are budgeting for a capital item, something that has a multiple year life and costs over $1000, you will need to keep in mind that you should include all of the costs to install or make the item useful.

If you are budgeting for the purchasing of a new piece of machinery, you want to determine an "all-in" cost. You will need to add in the indirect costs or overhead expenses, such as indirect labor (contract labor, for example), installation costs, maintenance, insurance, service contracts, and any other general expenses that will be needed to make the machinery serviceable. You will want to collect information from at least three vendors to determine the competitive price.

In preparing a budget for capital expenditures, you may want to start by making a list of all the property, plant, and equipment you will need to get the project completed. These items may include real property, machinery, equipment, vehicles, furniture, fixtures, and installations. Once you identify the items you need to acquire, you can start putting together cost information from catalogs, price lists, auctions, want ads, quotes, bids, offers, and appraisals. If the business involves building or remodeling a facility, a budget for the work could be prepared based on bids for the job.

The documentation you collect in putting together a capital expenditures budget will serve as good, solid support, and you may want to submit it to your accounting department together with your budget plan, and keep a copy for your own reference purposes.

Capital Budgeting Examples

1) Compactor
You would like to include a request in your capital budget for a compactor for drummed solid waste. Currently, your company generates four and a half drums of solid hazardous waste per month. You have contacted a compactor manufacturer and have gotten a written estimate for the compactor and shipping costs.

According to the manufacturer, you should be able to reduce the volume by the equivalent of three un-compacted drums to one compacted drum. You have contacted your facility's department and have gotten estimates on the cost of needed electrical service and floor support improvements. You have asked your accounting department for the appropriate discount rate (cost of capital). The following is the information, data, and assumptions.

Compactor FOB Cost	$27,500
Shipping Cost	$300
Electrical Upgrade	$2,500
Floor Upgrade	$750
Solid Waste Disposal Cost/Drum	$350
Allowed Drum Weight	450 lbs
Compaction Ratio	3 to 1
Cost of Capital	9%
Generation Rate/month	4.5 drums or 54 drums/year
Compactor Life	5 years

Calculations:

Drums per year = 4.5 x 12 = 54

Compacted drums per year = 54 / 3 = 18

Current disposal cost per year = 54 x $350 = $18,900

Future disposal cost per year = 18 x $350 = $6,300

Fully loaded cost of compactor = compactor cost + shipping + electrical & floor upgrade

= $27,500 + $300 + $2,500 + $750 = $31,050

Payback Period

Initial investment / yearly savings = $31,050 / ($18,900-$6,300)
=2.46 years
Net Present Value (NPV) is:

$$NPV = \sum_{t=1}^{n} \times \frac{CF_T}{(1+R)^t} - I_0$$

Where

NPV   = net present value
CF_T = sum of the annual cash flow
R   = required rate of return (Cost of Capital)
I_0 = net money invested (Initial Investment)

Let's show the calculation

$$NPV = \frac{12,600}{(1+.09)^1} + \frac{12,600}{(1+.09)^2} + \frac{12,600}{(1+.09)^3} + \frac{12,600}{(1+.09)^4} + \frac{12,600}{(1+.09)^5} - 31,050$$

$$NPV = 15,909 + 14,463 + 13,148 + 11,953 + 10,866 - 31,050$$

$$NPV = 49,010 - 31,050$$

$$NPV = \$17,960$$

So another way to think of the NPV is to say that the investment will produce a net return of $17,960 over 5 years considering the cost of money.

Internal Rate of Return

Let's use the same example, but this time we will calculate the investment's internal rate of return. Remember that the IRR is the rate that makes the projects NPV = zero.

Formula is:

$$I_0 = \sum_{t=1}^{n} x \ \frac{CF_t}{(1+IRR)^t}$$

So:

$$31,050 = \frac{12,600}{(1+IRR)^1} + \frac{12,600}{(1+IRR)^2} + \frac{12,600}{(1+IRR)^3} + \frac{12,600}{(1+IRR)^4} + \frac{12,600}{(1+IRR)^5}$$

$$IRR = 29\%$$

Now you have the data and calculations to present the business case for the purchase of the compactor. The payback period of 2.45 years is reasonable, the NPV is positive, and the IRR is a healthy 29%. In addition you may mention in your argument that the addition of the compactor will reduce transportation costs as well as the storage space needed, and will be a good waste minimization effort.

2) Learning Management Software

Your department has created computer based training courses. There are twelve courses that are sent to different groups ranging from the entire company to small groups of a few dozen. Currently your administrative assistant devotes 15% of her time to managing the system. The system is composed of a very large excel spreadsheet with all employees' names, positions, and the dates of different trainings completed. Your administrative assistant has done a good job of managing the system, but as the company grows she will need to invest more time to manage the system. You have researched learning management systems that may greatly improve the robustness of the system, reduce the likelihood of errors, and reduce the time necessary to administer the system. The learning management software is web delivered, and the quote that you have gotten is for a three year contract.

The following is the information, data, and assumptions:

Learning management software costs 1st yr.	$15,000
First year training and setup costs	$5,000
Percent of time admin. devotes to CBT	15%
Administrative assistant's yearly salary	$42,000
Cost of capital	10%

Investment

1st Year = $15,000 + $5,000 = $20,000

Savings
($42,000 x .15%) = $6300 / year

Payback Period

Savings per year = $6,300

$20,000 / $6300= 3.17 years

The payback period would exceed the term of the contract. You will not recoup the cost of the investment.

The formula for the NPV is:

$$NPV = \sum_{t=1}^{n} x \ \frac{CF_T}{(1+R)^t} - I_o$$

Where
NPV   = net present value
CF_T = sum of the annual cash flow
R   = required rate of return (Cost of Money)
I_o = net money invested (Initial Investment)

Let's show the calculation

$$NPV = \frac{6,300}{(1+.10)^1} + \frac{6,300}{(1+.10)^2} + \frac{6,300}{(1+.10)^3} - 20,000$$

$NPV = 5727 + 5206 + 4737 - 20,000$

$NPV = 15670 - 20,000$

$NPV = -\$4,330$

So another way to think of the NPV is to say that the investment will produce a net loss of $4,330 considering the cost of money.

For this example you will not be able to make a true business case for the purchase of the learning management software. You may look deeper to see if there maybe any additional savings. Perhaps HR or another department could use the software to help them with their training needs?

Chapter 10

From Data to Information

This chapter addresses one of the most important but difficult concepts to teach. With the advent of modern spreadsheet technology, we can all make graphs from data. Unfortunately, in my experience, many individuals' graphs are obscure. I have seen excellent data turned into un-intelligible graphs. This chapter should assist you in making better graphs by giving rules of thumb to apply when making graphs, displaying excellent and average graphs, and describing their strengths and weaknesses. Hopefully you can use the rules of thumb presented to avoid confusing graphs, and create graphs that clearly tell the story you want your audience to grasp.

At the end of the chapter I add a list of sources where you can find statistical & financial data and analysis tools that will assist you in developing your graphs and arguments.

Communicating Graphically

One of the most powerful tools for presenting the status of your department and showing future trends is by employing carefully crafted graphs. A well crafted graph should tell a story.

If you are presenting expense information, the y-axis should be denominated in dollars. You don't necessarily need to start the y-axis value at zero, but it is usually the most convenient. The scale of the y-axis should always exceed the largest value to be presented so that your graphs do not go "off scale". The x-axis normally represents time. I would encourage the manager to always present trends. Trends can only be shown if your graph's x-axis is expressed in time units whether it be years, quarters, or some other unit appropriate for your discussion.

A secret to making really expressive graphs is to have a y'-axis, that is a second y-axis to the right side of your graph. The purpose of the y'-axis is to be able to show the value of the trend line. The story of the graph is really told by the trend line. The trend line is the line that shows the unit cost, or the normalized cost for the information you are describing over time.

1. EH&S department costs
2. Department expenses as a percentage of total dept. costs
3. Hazardous waste costs
4. Medical exam costs
5. Worker's compensation insurance costs
6. Worker's compensation claims costs
7. Worker's compensation exposure modification factor
8. Hazardous waste generation
9. EH&S online trainings

Graph 1

Department Costs

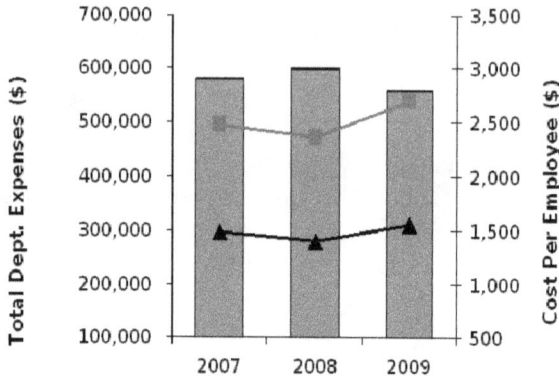

	2007	2008	2009
DEPARTMENT	581,000	600,000	560,000
ALL EMPLOYEES	1,486	1,405	1,551
SCIENCE ONLY	2,472	2,372	2,692

Please inspect the sample EH&S department cost in Graph 1 shown above. You can see by inspection that for the years 2007 through 2009 gross department costs have varied from approximately $600,000 to $560,000 per year. Furthermore, you can see that the normalized cost, which is the cost per employee per year, has varied very little and is approximately $1,500 per year. The cost per scientific employee per year has varied from approximately $2,500 to $2,700 during the three year period.

As you can see, I have chosen to use a bar chart to show the gross cost.

I hope by inspecting this graph that you can hear the story that is being told. That is, gross department costs have been reduced, but unit costs (as expressed as cost per scientist) have continued to rise modestly. This type of graphing allows you to present your overall department's cost status in a clear light. It allows for a discussion of what measures your department may pursue to control costs, and what future costs you expect. It allows you to clearly show executive management that you are tracking and trying to control required expenses.

Next, let's turn our attention to Graph 2 of EH&S department expenses as a percentage of total department costs. This graph's y-axis is in percents and the x-axis is in years. Instead of one trend line, we have a few. This allows you to rank different trend lines. I suggest that you never exceed four trend lines on any one graph, as an excessive number of lines obscure the graph. Graph 2.1 may be an excellent way of showing the average percentage costs.

Graph 2

Department Expenses as a Percentage of Total Department Costs

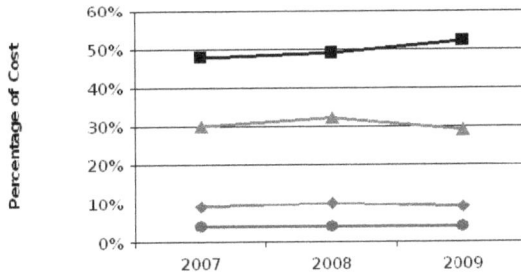

	2007	2008	2009
SALARIES	48%	49%	52%
LAB SERVICES	30%	32%	29%
SUPPLIES, GASSES, DRY ICE	9%	10%	9%
RENTAL AND REPAIR	4%	4%	4%

Graph 2.1

Department Expenses as a Percentage of Total Department Costs
Average Expenses from 2007-2009

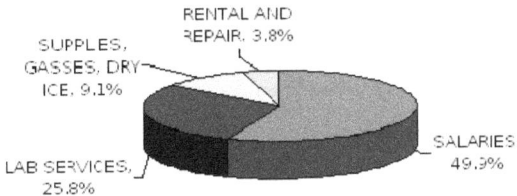

RENTAL AND REPAIR, 3.8%

SUPPLES, GASSES, DRY ICE, 9.1%

SALARIES, 49.3%

LAB SERVICES, 25.8%

By inspection, the graph allows you to determine those expenses that make up the largest percentage of your department. It also allows you to communicate clearly to executive management and clearly identify those expenses that could be reduced if the situation required cost reductions.

For example, if you were mandated to reduce your budget by x dollars, it may be difficult to find the dollars in your rental and repair budget. However, it may be possible to identify some way to reduce costs in your lab services expense budget since it makes up a much larger percentage of your overall department costs.

Graph 3

Hazardous Waste Costs

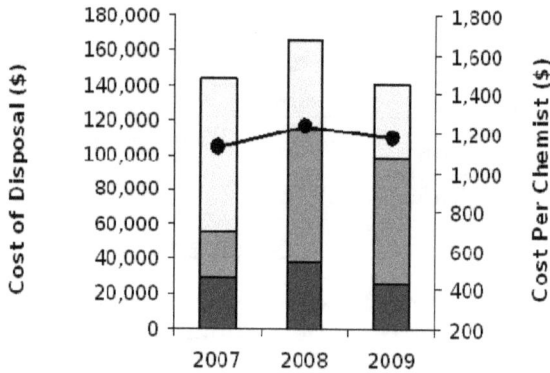

	2007	2008	2009
CH	87,000	52,000	42,000
VEOLIA	27,000	76,000	73,000
AET	29,000	38,000	26,000
COST/CHEMIST	1,126	1,230	1.175

Please inspect the hazardous waste cost in Graph 3 above. You can see by inspection that for the years 2007 through 2009, gross hazardous waste costs have varied from approximately $140,000 to $165,000 per year. Furthermore, you can see that the normalized cost, which is the cost per chemist per year for hazardous waste disposal, has varied from approximately $1,000 to $1,300 per year during the three year period.

Now comes more discretionary presentation of data. As you can see, I have chosen to include stacked bars to show the gross cost for each hazardous waste disposal vendor. Some may argue that I have presented too much data in one graph. You must use your own discretion, but if you are to err it is usually better to present less information on one graph than to overload it. You want your graph to tell a clear story. Remember that clutter obscures the story.

I hope by inspecting this graph, you can hear the story that's being told. That is, gross hazardous waste costs have leveled off. Unit costs have continued to rise modestly but appear to be leveling off.

This type of graphing allows you to present your department's cost status for a particular expense in a clear light. It allows for a discussion of what measures your department may pursue to control costs, what future values you expect for the expense by extrapolation, and it allows you to clearly show executive management that you are tracking and trying to control required expenses.

Graph 4

Medical Exam Costs

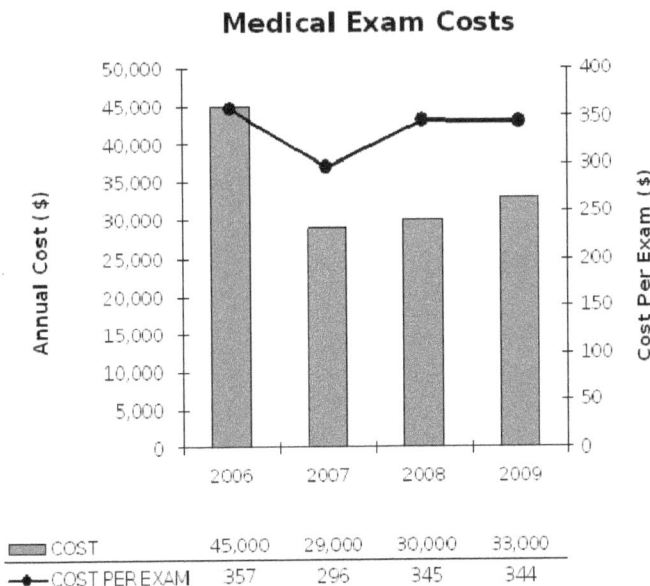

	2006	2007	2008	2009
COST	45,000	29,000	30,000	33,000
COST PER EXAM	357	296	345	344

Please inspect the Medical Exams costs in Graph 4 above. You can see by inspection that for the years 2006 through 2009, gross medical exam costs have varied from approximately $60,000 to $35,000 per year, and appear to be leveling off. Furthermore, you can see the normalized cost, which is the cost per employee per year, has varied very little after 2007 and is approximately $350 per year.

As you can see, I have chosen to use a bar chart to show the gross cost.

By inspection of this graph, I'm sure that you can hear the story that's being told. That is, gross medical costs have been generally reduced, and are now plateaued at about $37,500 per year, and the per employee unit cost, expressed as cost per employee's exam, has also stabilized at about $350 per year.

Graph 5

Workers' Compensation Insurance Costs

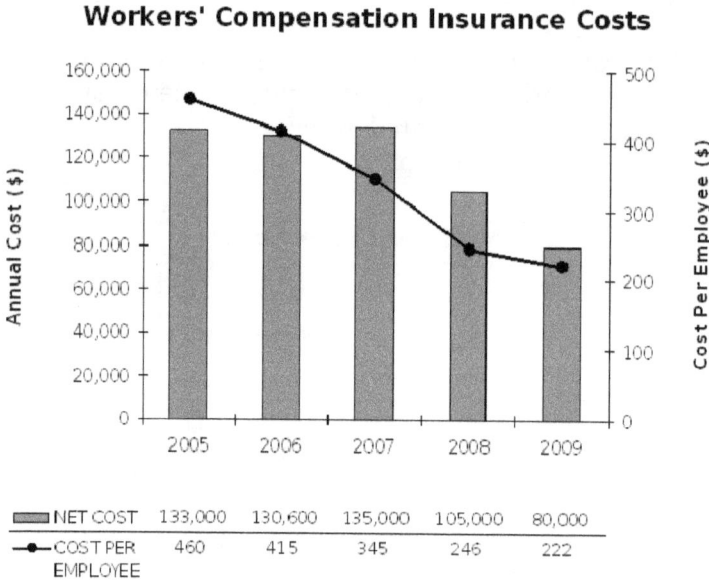

	2005	2006	2007	2008	2009
NET COST	133,000	130,600	135,000	105,000	80,000
COST PER EMPLOYEE	460	415	345	246	222

Please inspect the Workers Compensation Insurance costs in Graph 5 above. This would be a graph that any EH&S manager would like to present. You can see by inspection that for the years 2005 through 2007, gross insurance cost plateaued at about $134,000 per year. Thereafter, in 2008 & 2009, we see substantial cost reductions.

Furthermore, you can see the normalized cost (trend line), the cost per employee per year, has followed the exact trend noted above, but cost were reduced by almost 50% per year during the five year period. The story of this graph is that you are doing something right! That being gross & normalized workers compensation insurance costs have been significantly reduced!

This type of graphing allows you to present your overall workers compensation insurance savings very clearly. It allows you to demonstrate to executive management that your leadership in this area is working.

Graph 6

Workers' Compensation Claims Costs

	2005	2006	2007	2008	2009
COST OF CLAIMS	5,709	1,734	6,016	7,290	6,216
AVERAGE COST/CLAIM	519	289	752	810	1,036

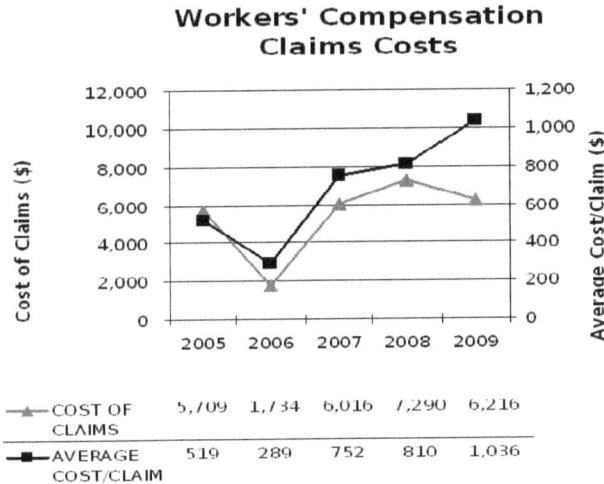

You may want to follow up with an example of Graph 6: workers compensation claims cost trends. This graph's y-axis is in dollars and the x-axis is in years. The y-prime is also in dollars per average claim. Instead of one trend line, we have two. These trend lines together tell a story. The story is that the total cost of claims has generally been consistent over the period at around $5,000 with a onetime drop in 2006.

Two other facts appear: (1) the overall number of accidents has been reduced but the cost per accident has been increasing, and (2) the cost per employee per year has varied very little and is approximately $1,500 per year.

Graph 7

Workers' Compensation Experience Modification Factor

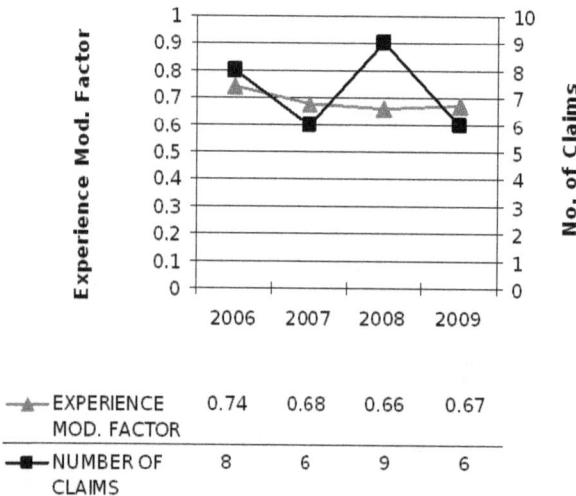

	2006	2007	2008	2009
▲ EXPERIENCE MOD. FACTOR	0.74	0.68	0.66	0.67
■ NUMBER OF CLAIMS	8	6	9	6

Next let's turn our attention to Graph 7, a graph of workers compensation experience rating. As you know, the workers compensation experience rating is a unit-less number. A new firm starts out with a rating of 1.00. The National Council on Compensation Insurance, Inc. (NCCI) creates the experience modification rating (EMR) for your firm based on your industry, classification of workers, and your company's loss history. If you have tried to investigate the formula, as I have, you will find it is their trade secret. Theoretically the EMR can be used to rank your performance over time and relative to other organizations.

This graph's y-axis is in a unit-less number, the x-axis is in years and y' axis is in number of claims per year. Instead of one trend line, we have two. One trend line shows the number of claims per year and the second shows the EM trend line.

The story this graph tells is that you are doing an OK job. That is, the EM trend line has tended to decline with time and now appears to be leveling off at a relatively low level. The number of workers compensation claims has been within a reasonable band.

It may have been better to present the number of claims normalized to claims per 100 employees, which is the standard index. That would have told a clearer story. Since we don't know if the number of employees has increased, decreased, or stayed the same, we are unable to judge the true performance.

Graph 8 is not a financial graph but one of environmental performance.

Graph 8

Hazardous Waste Generation

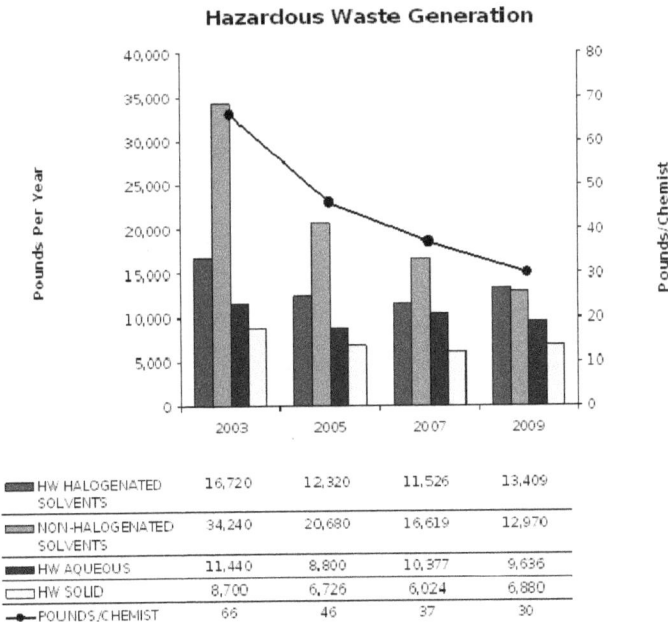

	2003	2005	2007	2009
HW HALOGENATED SOLVENTS	16,720	12,320	11,526	13,409
NON-HALOGENATED SOLVENTS	34,240	20,680	16,619	12,970
HW AQUEOUS	11,440	8,800	10,377	9,636
HW SOLID	8,700	6,726	6,024	6,880
POUNDS/CHEMIST	66	46	37	30

Graph 8, a graph of hazardous waste disposal for a company over the years 2003 to 2009, presents the quantity reduction in this case by waste stream type. The y axis is in pounds, the x axis is in years and the y' axis is in pounds per employee per year.

This graph is a bit busy, with multiple bars and a trend line, but it does show clearly that the amount of hazardous waste in gross terms and normalized terms (trend line)has been steady reduced over the time period. The story this graph tells is that you are again doing something right! That is, both gross & normalized quantity of hazardous waste has been reduced.

Graph 9 displays the number of online trainings presented over the period of 2005 to 2009, and is relatively straightforward. The story told is that there is a trend for an increasing number of online trainings, with a slight drop in 2009. You may want a graph like this to argue that you need more resources to improve your online trainings.

Graph 9

EH&S Online Training

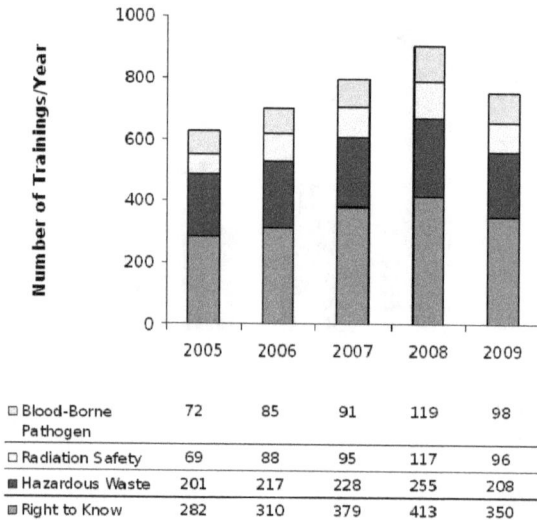

	2005	2006	2007	2008	2009
☐ Blood-Borne Pathogen	72	85	91	119	98
☐ Radiation Safety	69	88	95	117	96
■ Hazardous Waste	201	217	228	255	208
▨ Right to Know	282	310	379	413	350

Graph 9.1, a pie chart, may be a very useful tool to clearly show the relative frequency of a course. I believe that a superior presentation may be made by using the two graphs. It may be better to show the bar chart with only total courses and using the pie chart to describe the relative course frequency.

Graph 9.1

**EH&S Online Training
Number of Trainings 2005-2009**

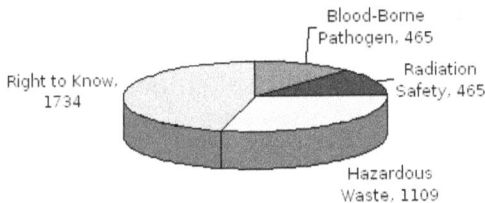

I have not exhausted the possible graphs that you may want to create. But to make graphs you need data and an idea of what information you want to present. I am providing some sources for data and analysis. The list is by no means exhaustive, but hopefully it will put you on a productive path.

I strongly recommend that you attend one of Robert Emery's professional development courses on graphing usually held at the AIHce Conference. Robert Emery, PhD, is the Vice President for Safety, Health, Environment, and Risk Management & Quality Assurance for The University of Texas Health Science Center.

Sources for Data & Analysis Tools

The Bureau of Labor Statistics (BLS)[27]

The Bureau of Labor Statistics (BLS) is a unit of the United States Department of Labor. It is the principal fact-finding agency for the U.S. government in the broad field of labor economics and statistics. The BLS is a governmental statistical agency that collects, processes, analyzes, and disseminates essential statistical data to the American public, the U.S. Congress, other Federal agencies, State and local governments, business, and labor representatives. The BLS also serves as a statistical resource to the Department of Labor[28].

Council for State and Territorial Epidemiologist (CSTE)[29]

The Council of State and Territorial Epidemiologists (CSTE) was organized in the USA in the early 1950s in response to the need to have at least one person in each state and territory responsible for public health surveillance of diseases and conditions of public health significance. Since then, CSTE has grown to include members from every U.S. state and territory, Canada, and Great Britain.

The surveillance and epidemiology of infectious diseases, chronic diseases and conditions, and environmental health concerns are priority areas for CSTE. Members serve as special topic consultants for a broad range of public health concerns such as HIV/AIDS and vaccine-preventable diseases[30].

The Mountain & Plains Education and Research Center[31]

The Mountain & Plains Education and Research Center is one of seventeen Education and Research Centers funded by the National Institute for Occupational Safety and Health (NIOSH). NIOSH and the NIOSH Education and Research Centers are affiliated with the Centers for Disease Control and Prevention.

The Mountain & Plains Education and Research Center is collaboration between the University of Colorado Denver (Anschutz Medical Campus), Colorado State University, National Jewish Health, the University of New Mexico and Denver Health.

The Occupational Health Indicators in Colorado: A Baseline Health Assessment (2001-2005) has excellent summaries of information for the state of Colorado. The report can be found at MAP ERC or http://www.cdphe.state.co.us/hs/adultdata/OHI%20Colorado%202001-2005.pdf

The Global Reporting Initiative (GRI)[32]

The Global Reporting Initiative (GRI) produces one of the world's most prevalent standards for sustainability reporting - also known as ecological footprint reporting, Environmental Social Governance (ESG) reporting, Triple Bottom Line (TBL) reporting, Corporate Social Responsibility (CSR) reporting. Sustainability reporting is a form of value reporting where an organization publicly communicates their economic, environmental, and social performance. GRI seeks to make sustainability reporting by all organizations as routine as, and comparable to, financial reporting.

GRI Guidelines are regarded to be widely used. As of January 2009, more than 1,500 organizations [1] from 60 countries use the Guidelines to produce their sustainability reports. (View the world's reporters at the GRI Reports database). GRI Guidelines apply to corporate businesses, public agencies, smaller enterprises, NGOs, industry groups and others. For municipal governments, they have generally been subsumed by similar guidelines from the UN ICLEI[33].

Puget Sound Chapter of the Human Factors and Ergonomics Society[34]

To help practitioners better quantify the benefits of ergonomics, Rick Goggins, a Labor and Industries ergonomist and PSHFES Council Member, has developed a cost-benefit calculator that PSHFES is proud to offer for free download.

The calculator allows you to compare three intervention options, and provides estimates of benefits and payback periods. The calculator is based on a review of 250 case studies in which organizations reported the outcomes of ergonomics programs and individual solutions.

University of Texas Health Science Center at Houston (UTHealth) Safety, Health, Environment, & Risk Management[35]

Indicators of Performance in the Areas of Losses, Compliance, Finances, and Client Satisfaction

The objective of this report is to provide a metrics-based review of safety, health, environmental, resource management (SHERM) operations

Chapter 11

Risk Management

Risk Management is a critical aspect of all EH&S managers' duties. We all have a sense of what the goals are, identifying hazards and their likelihood, and then designing administrative and engineering solutions to reduce or mitigate the hazard. I expect that all EH&S managers have had to identify hazards and reduce the risk of those hazards. Unfortunately, I do not believe we as a group are as adept at estimating the financial cost of mitigation. For mature organizations, we need to address business continuity. As the complexity of the challenge increases, the estimation of financial risks becomes more difficult. Luckily with larger organizations, the manager will have more resources to draw upon to estimate financial costs of risk mitigation.

No matter how small your organization is, it should be aware of its risk situation and take appropriate measures to protect itself against operational and financial exposures. All EH&S managers should know and practice risk management. Risk management strategies should be integrated into the day-to-day functions of the organization.

A Risk Management Program is a systematic effort to reduce risk. It begins with a formal, written risk management plan that is clearly communicated to your organization's employees. They must understand its basic elements and goals.

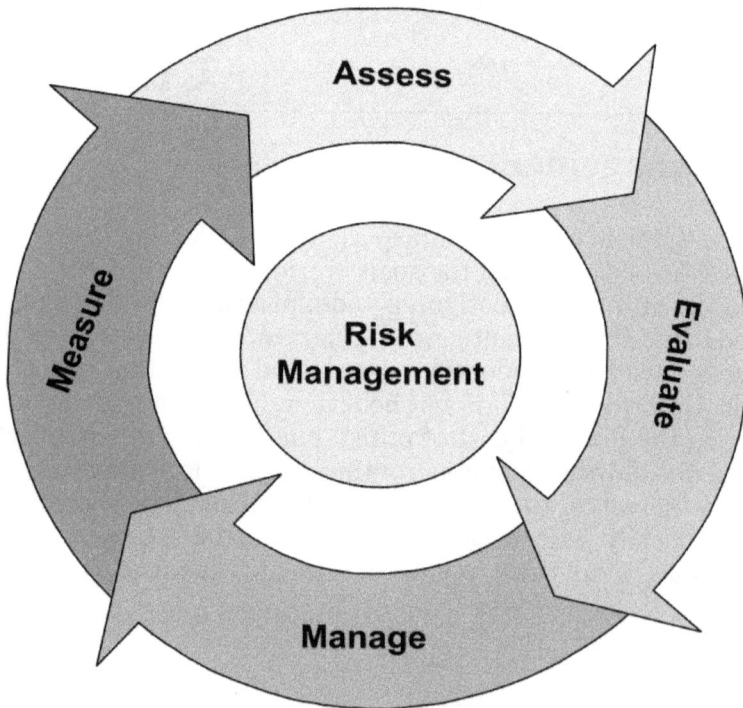

A Risk Management Program will require you to research your specific hazards. That is not to say you should not use other resources. You should not reinvent the wheel, but use available, industry-wide resources when available. The risk management program must be able to collect and analyze pertinent data. This information is then analyzed with the intent of improving your organization's risk posture.

Identifying Risks

You cannot manage risk until you have a system in place for identifying and tracking unusual events (incidents) that occur in your facility. The most common method for keeping tabs on incidents is the incident report form. This is a standardized format that enables staff members to document in a clear, concise, and consistent manner any events they witness that deviate from routine care.

The form should capture relevant, objective information regarding the event and surrounding circumstances. It also expedites notification of management for worker compensation and accident reporting purposes. The form should facilitate the entry of data into a computer database designed to track and trend events. By analyzing data according to type of event, time of day, unit, and department, you can identify where and when problems tend to originate. Incidents should be categorized both by frequency and by severity. Your organization directives on the use and requirement of completing an incident report should be clear. All the organization's members should know when and how to complete an incident report form.

Incidents are not complaints, but both require similar responses. In each case a response should be to the individual filing the incident and their supervisor. ***The important thing is to clearly identify the root cause and insert the correction.*** Hopefully you will be able to identify areas requiring improvement. Sometimes your findings are so important for the safety of the organization that they should be distilled and broadcast to the organization. Be careful not to include any individually identifying information to avoid embarrassing incident reporters.

Your risk management program should address the following issues:

- To what individual or department is the form to be sent?
- Who will take responsibility for responding to the incident or complaint?
- Who is responsible for monitoring follow-up and resolution of incidents/complaints?

The key objectives of the risk management program should be:

- Systematizing risk management
- Preventing accidents/injuries/adverse incidents
- Inserting needed corrections/remedial actions
- Documenting incidents
- Educating the organization on requirements and important revelations

Additional References:

To learn more about the principles of risk management, consult:

The Risk Management Handbook, from the American Society for Healthcare Risk Management [36]

DRAFT INTERNATIONAL STANDARD ISO/DIS 31000[37]

Industrial Hygiene & Safety Auditing - A Manual for Practice, Second Edition[38]

Risk Assessment Principles for the Industrial Hygienist[39]

In the next section, we will see that business continuity is an extension or a special case of risk management. The focus will shift from employee accidents and incidents to larger unforeseen events that could trigger significant disruption for the entire organization.

Business Continuity

What is business continuity? I believe that there are many useful definitions, but I believe the following definitions[40] best illuminate facets of the practice.

Business continuity is:

>the ability to maintain operations/services in the face of a disruptive event

>the ability of an organization to continue to function even after a disastrous event, accomplished through the deployment of redundant hardware and software, the use of fault tolerant systems, as well as a solid backup and recovery strategy

>the activity performed by an organization to ensure that critical business functions will be available to customers, suppliers, regulators, and other entities that must have access to those functions

As you review the definitions above, you will see that the perspective changes. The first, which is the clearest, speaks about an organization maintaining its ability to function after a disruptive event. The second definition is from an information technology perspective. The third definition identifies critical business functions in a larger organization being provided to *customers and outside entities* in spite of a disruptive event.

Another way to approach the question is to ask what are some of the organization's most valuable assets?

- Intellectual property
- Financial assets
- Tangible property
- Reputation

How would you protect them in the event of a significant unplanned event?

I believe that business continuity planning is a reflection of an organization's maturity, size, and critical functions. If your business is a small start up operation, then business continuity means keeping the doors open and the staff paid. The organization may not have the luxury of time to plan for the future; hope and prayer will have to suffice.

If a company has matured to the state where it is stable and has reasonable retained earnings, then business continuity may mean insuring and protecting the business from a limited number of foreseeable events, but not events that necessarily affect community-supplied services. For instance, a manufacturing company may have backup power to tide critical systems over during short power outages, but few companies would consider having sufficient backup power to keep manufacturing their product in the face of a power outage.

If the organization is a blood bank, then *you know* that your product will be needed at critical times. *If you have not planned for the protection and delivery of your critical product (service) during disruptive events then you have failed as an organization.*

Business Continuity & EH&S Management

It is natural for some of the functions of business continuity to fall on the EH&S manager. He or she should be well versed in risk avoidance, planning for natural disasters, and mitigation of incidents and crises management.

Disruptive events may take many forms. While natural disasters such as flooding, hurricanes, or earthquakes may be infrequent events, more common causes of disruptive events can strike at any time, from power outages to computer viruses to disruption by discontented employees.

In December 2006, the British Standards Institution (BSI) released a standard for business continuity planning [41](BCP), BS 25999-1. The standard's applicability was designed to extend to organizations of all types, sizes, and missions, whether governmental or private, profit or non-profit, large or small, or industry sector.

In 2007, the BSI published the second part, BS 25999-2 "Specification for Business Continuity Management", which specifies requirements for implementing, operating, and improving a documented Business Continuity Management System (BCMS).

These UK standards maybe a useful starting point for EH&S managers to approach the task of business continuity planning.

Many organizations and companies aren't adequately prepared for significant disruptive events. I believe that the greatest barrier to preparation is a lack of executive support and concomitant funding. Adequate funding, executive management support, and a capable designated leader are prerequisites for a successful business continuity effort.

I believe that an iterative approach would be the best way to pursue the overall objective. That is to start small, develop your team's expertise, select one functional area to concentrate on, learn from your successes and challenges, and then expand the effort to a larger section of the organization. These steps could be conducted sequentially or concurrently for different business functions. For example the first area to be tackled may be information technology

(IT) and computer systems. An intelligent approach may be to find a champion in the IT department for your team to work with for this business unit.

Once the team has become more adept at addressing the issues at one IT location, the same approach could be applied to all of the organization's IT locations. If you had a large enough team, multiple business functions could be evaluated simultaneously. However, remember that one size may not fit all situations. For example, the electrical power supplied in advanced countries is relatively stable and secure; that may not be true in less developed nations.

What are reasonable management steps to pursue?

- Manager selects and empowers the business continuity leader
- Leader assembles the initial team
- Team researches applicable published guidance and similar organizations' best practices, and possibly employs consulting experts
- Team decides on the most fruitful approach and tools to employ
- Leader presents selected approach for executive management's review
- Leader selects one of the organization's functions to concentrate on and apply approach
- Team uses lessons learned to modify/improve approach prior to full roll out
- Team produces metrics to measure improvement and reinforce accountability
- Team produces the draft business continuity plan
- Team creates applicable training documents
- Team enrolls selected business unit managers to assist and participate in the larger pan-organizational effort
- Team tests completed plan in intellectual thought experiments
- Team creates and disseminates training for the general organization
- Team tests the plan in simulated situations
- Leader continually updates plan to incorporate organizational changes

The above was not intended to delineate the nuts and bolts of how to do business continuity planning. Nut-and-bolt tasks usually incorporate the following:

- Identifying and ranking hazards/risks
- Analyzing and defining requirements for recovery
- Designing and documenting the business continuity plan and recovery operations
- Conducting implementation training
- Testing the business recovery plan
- Maintaining and updating the business recovery plan

Tasks for a business continuity manager of a large, global corporation

To be successful, a large, global corporation business continuity manager would need a mandate from senior management. Without an explicit directive to cooperate with the manager, the inertia of a larger business would stop the effort in its tracks. The manager would need to assemble a team to develop the company business continuity function globally.

Responsibilities for initiatives must originate or be synthesized by the team, but the actual effort would be companywide. The first objective should be to establish relationships with those in the different organizational units who know what risks are present in their areas. Next would be the development of strategies and systems to analyze risk and design potential mitigation actions. Again, each organization will have a different set of needs and priorities. Below is a suggested list of responsibilities for the business continuity manager of a large global organization,

Manager Responsibilities for a Global Corporation:

> Develop, implement, and continually improve a corporate EH&S business continuity program to address financial, operational, compliance, and reputational risks.

Develop strategies with each facility functional unit to incorporate specific action plans.

Develop benchmarks with internal and external organizations to catalog best-in-class practices to support the risk mitigation and management process.

Enroll organizational leaders to secure alignment, consistency, and superior results.

Build relationships with key departments to foster support for continual improvement.

Develop strategies to monitor and determine impacts of current and emerging EHS regulations.

Establish action plans to respond to new regulations which could have an impact on company operations.

Create leading and lagging tracking metrics to drive improvement and reinforce accountability.

Supervise the development and implementation of continuity training programs.

Develop tools for business units to establish effective self-audit processes.

Lead internal corporate compliance investigations.

Additional References:

You may find the following books to be a significant resource in your effort to establish a corporate business continuity program:

Crisis Communication: Guidelines for Action Planning[42] (DVD)

Investigations: A Handbook for Prevention Professionals[43]

Environmental Health and Safety Audits[44]

Business Continuity: Best Practices-World-Class Business Continuity Management, Second Edition[45]

Business Continuity Planning: A Step-by-Step Guide with Planning Forms on CD-ROM, Third Edition[46]

The Definitive Handbook of Business Continuity Management[47]

Chapter 12

Preparing Your Presentation

In this chapter we will revisit the original challenge of getting your presentation ready to deliver to executive management. We will review the objectives of your presentation, and suggest verbiage and graphical content to include. The next chapter will turn the suggestions into a PowerPoint presentation.

Assume that you will be misunderstood even if you are clear. Make your presentation crystal clear. What cannot be clearly described should not be considered. Otherwise, your meeting will be diverted into some tangential discussion. Let's refresh our memories on what we are planning to present to executive management.

1. You understand executive management's goals and are striving to align your department to execute them.

2. You have studied your company's operations. You understand exactly what business your company conducts and the body of laws, regulations, and best practices with which your company will need to align itself.

3. You are managing your staff effectively. You have:
   a. Determined staffing requirements.
   b. Hired or are in the process of hiring capable staff.
   c. Established a system to supervise direct reports to maximize their effectiveness.
   d. Set your employees annual goals and objectives.
   e. Delegated required tasks to staff.
   f. Planned and implemented procedures and systems to maximize staff efficiency.
   g. Instituted an EH&S management system to ensure task completion.

4. You are managing your resources effectively. You have:
   a. Employed financial and budget tools to maximize operational efficiency.
   b. Facilitated the preparation of appropriate management reports.
   c. Reviewed performance data and reports to monitor and measure productivity.

Okay, let's address each and every major objective and discuss how you may use the insights and tools presented to reach your goal.

You could start with the simple thing: that is, state that your department's goal. For example, "While conserving financial resources, provide excellent risk management & EH&S services to the organization." This would be an excellent time to present a graph of your department's gross and per employee costs for the last few years.

Next you may add some objective measurements:

For instance, you could state that your goal is to contain year over year normalized expenses growth for your department to below 5%, and cooperate with the corporate insurance company to conduct a facility wide audit within the next 12 months to identify potential compliance weaknesses. A service goal may be to respond to employee requests for assistance within 2 business days.

You don't want to bore them to tears by citing chapter and verse of the company's operations, body of laws, regulations, and best practices that your company will need to align itself with. Rather, state a summary and then choose a few salient points to develop.

"After a review of the company's operations and state, federal, and local requirements, I have determined that we have 65 significant recurring tasks that must be completed annually. Tasks have been assigned to appropriate staff and company departments. I review the system weekly to ensure task completion."

You may also describe future needs that you see on the horizon such as "I believe that we will need to add a Potent Compound Control Program this year to compliment the expansion of our cGMP capabilities".

You engage executive management by having a discussion of who you may work with inside the company to facilitate the creation of the program. If you do start a conversation, make sure that it does not get sidetracked.

One way you may summarize the effectiveness and completeness of your staff utilization is to describe coverage areas and management systems. For example: "PM fulfills our hazardous waste management's needs at one third the cost of a contracted service, while also assisting with safety and industrial hygiene services." "VS has mastered our compliance record keeping, updates the company's EH&S intranet, and services as a communications point for internal & external questions, requests, and general assistance." " HM serves as the contact point in our Longmont facility, and is responsible for program updates."

We have an EH&S management system in place that catalogs all of the identified recurring tasks, identifies the responsible party for the task, and documents the completion of the task. We meet monthly to review the status of evolving programs and ideas for improving department services. We have all of the needed staff in place, and each member has been assigned performance goals.

In describing your management of resources you should show your tracking of expenses. This would be an excellent time to present some of the graphs you have created. I would present a global picture by presenting the EH&S department cost graph if you have not presented it earlier, EH&S department expenses as a percentage of total department costs, and a graph of the next most costly department expense. This would also be the appropriate time to describe and possibly show how you are tracking expenses for budgeting purposes, and describe how you are interfacing with the accounting department. If you have the ability, you should show budgeted expenses vs. actual expenses.

Chapter 13

Briefing Executive Management

This chapter presents a PowerPoint example that is designed to be clear and targeted to executive management. Obviously it can only be a general example. You will need to work to make it specific to your organization.

You should have practiced giving your *full* presentation at least *three* times prior to the actually delivery. Time yourself after you have established a smooth flow to the presentation. Make sure that you are not rushing through the presentation. Use a measured pace. Build in time for questions. I suggest you have a trusted member of your staff evaluate your presentation, both for content and delivery. Don't leave things to chance. Be prepared so that you will be confident when you present. Confidence is key!

Choose an appropriate conference room for your presentation, preferably one the executives frequently use. Book the room at least two months in advance. Set a date that is between quarters so executive management will not be rushing to finish quarterly reports. Executives often go on the road to promote the company; ask one of the executive assistants to suggest a time when they believe that most of the executives will be in town.

Have your supporting documents ready to access electronically & project. If you will be including individuals in a remote broadcast *make sure that that the remote communication system is working*. Test it for the room you will be using. Test all of the systems at least 1 hour before your presentation. Know who in IT can help you if you need assistance. Nothing is more frustrating that a dead battery on a pointer or keypad. Your presentation should be about your work. Make sure that all eccentricities are handled.

EH&S Department Presentation to Management

January 30, 2010

By James L Lieberman

Department Goals

The EH&S department's goal is to provide excellent risk management & EH&S services to the organization while conserving financial resources by:

- Aligning the department to support corporate goals

- Being a subject area expert resource for the company

- Incorporating succinctly the requirements from laws, regulations, and best practices into the company's guidance documents and policies

- Managing department staff effectively

- Managing the departments resources effectively

Service Goals for 2011

- Respond to employee requests for assistance within 2 business days
- Respond within 1 day to all accidents / worker compensation incidents
- Be the lead department for business continuity planning efforts

Management Systems

After a review of the company's operations and state, federal, & local requirements, we have determined that we have 65 significant recurring tasks that must be completed annually

We have instituted a EH&S management system to systemize task tracking

Recurring tasks have been assigned to appropriate staff and company departments

Task completion is reviewed weekly to ensure task completion

Proposed New Program for 2011

To support the expansion of our cGMP capabilities to manufacture Tox Lot quantities of API we propose to:

Create a Potent Compound Control Program

The objectives of this program will be to:

- Protect cGMP personnel from overexposure to API
- Set performance standards for cGMP equipment
- Provide and document API hazard communication to Process Chemists
- Provide a tool to categorize API within a banding system
- Provide a tool to communicate API hazards to contract manufacturing
- Organizations

To be successful we will need executive management's full support.

Fiscal Goals for 2010

- Contain year over year normalized EH&S expense growth below 4%.
- Reduce workers compensation costs by 5% by becoming cost containment certified by the State Department of Labor

Department Costs

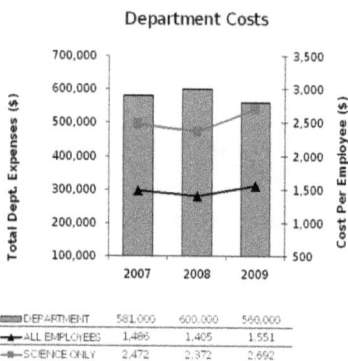

	2007	2008	2009
DEPARTMENT	581,000	600,000	560,000
ALL EMPLOYEES	1,486	1,405	1,551
SCIENCE ONLY	2,472	2,372	2,692

Department Expense Trends

The graph below highlights the department's four largest expenses

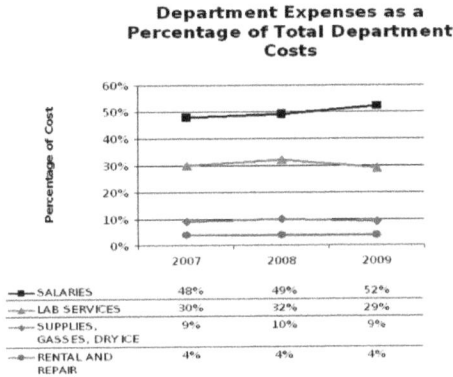

Department Expenses as a Percentage of Total Department Costs

	2007	2008	2009
SALARIES	48%	49%	52%
LAB SERVICES	30%	32%	29%
SUPPLIES, GASSES, DRY ICE	9%	10%	9%
RENTAL AND REPAIR	4%	4%	4%

Expense Reduction & Environmental Sustainability

The department has worked diligently for the past four years to reduce hazardous waste quantities and disposal costs

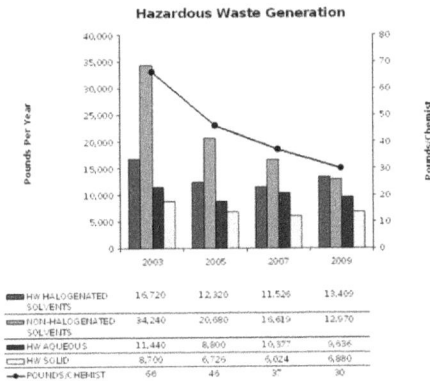

Hazardous Waste Generation

	2003	2005	2007	2009
HW HALOGENATED SOLVENTS	16,720	12,320	11,526	13,409
NON-HALOGENATED SOLVENTS	34,240	20,080	16,619	12,970
HW AQUEOUS	11,440	8,800	10,377	9,636
HW SOLID	8,700	6,726	6,024	6,880
POUNDS/CHEMIST	66	46	37	30

Expense Reduction & Environmental Sustainability

Hazardous Waste Costs

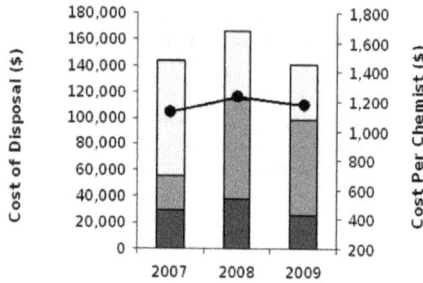

	2007	2008	2009
CH	87,000	52,000	42,000
VEOLIA	27,000	76,000	73,000
AET	29,000	38,000	26,000
COST/CHEMIST	1,126	1,230	1,175

Expense Reduction for Worker's Compensation Insurance

We have worked successfully to reduce workers compensation insurance costs

Workers' Compensation Insurance Costs

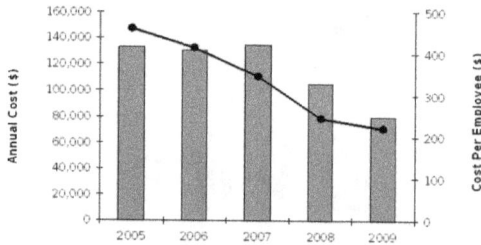

	2005	2006	2007	2008	2009
NET COST	133,000	130,600	135,000	105,000	80,000
COST PER EMPLOYEE	460	415	345	246	222

Capital Item Request

We are requesting authorization for an expenditure of $ 62,100
for two waste compaction systems to reduce annual disposal costs
from $37,800 to $ 12,600

Payback period: ~ 2.5 years

NPV $35,920

IRR   29%

We would be happy to entertain questions or comments.

About the Author

James Lieberman, CIH

James is the president of Environmental Information Services, Inc. (EIS) in Boulder, Colorado. He helps EIS' clients increase their compliance, responsiveness to environmental issues and profitability. Clients included federal and state governments, local and international companies. He has served as an expert witness nationally. Before that he served as the Manager of EH&S at Array BioPharma, a Colorado-based drug discovery company, for nine years.

He is an author of <u>A Practical Guide for Hazardous Waste Management, Administration and Compliance</u>. His areas of specialty include comprehensive industrial hygiene services, assisting clients with program management and cost containment. His focus is to support teams in implementing programs that provide speed, flexibility, and efficiency for industrial operations and research and development, while integrating EH&S into all portions of the business.

He has taught management and safety courses to both private and public groups, and was honored by Colorado Mtn. College with an award for instructor of the year in 1984. He has been an active member of the American Industrial Hygiene Association since 1990. In 2009 he was honored by the local section as Industrial Hygienist of the year.

He has a B.S. degree in chemistry from the University of Richmond, and an M.B.A. with an emphasis in finance from the University of Colorado. He is certified by the American Board of Industrial Hygiene for the comprehensive practice of industrial hygiene.

References

[1] Bartley, WW. (1988). The Transformation of Man: The Founding of EST. New York: Clarkson Potter.

[2] Savitz, A.W., & Weber, K. (2006). The Triple Bottom Line: How Today's Best-Run Companies are Achieving Economic, Social and Environmental Success -- and How You Can Too. San Francisco, CA: Jossey-Bass.

[3] Creative Research Systems, . (2007). Survey design. Retrieved from http://www.surveysystem.com/sdesign.htm

[4] SurveyMonkey. (1999). Retrieved from http://www.surveymonkey.com/

[5] Geller, E.S., & Toms, L.A. (1998). People Based Safety: The Source. Coastal Training Technologies Corp.

[6] Maslow's Hierarchy Of Needs. (2010). Wikipedia. Retrieved September 25, 2010, from http://en.wikipedia.org/wiki/Maslow%27s_hierarchy_of_needs

[7] Maslow, A.H. (1943). A Theory of Human Motivation. Psychological Review, 50, 370-396.

[8] Savitz, A.W., & Weber, K. (2006). The Triple Bottom Line: How Today's Best-Run Companies are Achieving Economic, Social and Environmental Success -- and How You Can Too. San Francisco, CA: Jossey-Bass.

[9] Wiesel, E. (n.d.). A Templeton Conversation: Does the Universe Have a Purpose?. Retrieved from http://www.templeton.org/purpose/essay_Wiesel.html

[10] Geller, E.S., & Toms, L.A. (1998). People Based Safety: The Source. Coastal Training Technologies Corp.

[11] Collins, J.C., & Porras, J.I. (1997). Built to Last: Successful Habits of Visionary Companies. New York, NY: HarperCollins Publishers.

[12] Lombardo, M.M., & Eichinger, R.W. (2004). FYI: For Your Improvement, A Guide for Development and Coaching. Lominger International.

[13] Kouzes, J.M., & Posner, B.Z. (2007). The Leadership Challenge. San Francisco, CA: Jossey-Bass.

[14] Collins, J. (2001). Good to Great: Why Some Companies Make the Leap. And Others Don't. New York, NY: HarperBusiness.

[15] American Management Association, . (n.d.). American Management Association: Seminars. Retrieved from http://www.amanet.org/individualsolutions/seminars.aspx?SelectedSolutionType=Seminars

[16] Pryor, Fred. (1999). Fred Pryor Seminars & CareerTrack. Retrieved from http://www.careertrack.com/site/default.aspx

[17] AIHA Management Committee. (2001). Industrial Hygiene Performance Metrics Manual. Fairfax, VA.

[18] International Organization for Standardization. (2010). International Standards for Business, Government and Society. Retrieved from http://www.iso.org/iso/home.html

[19] International Organization for Standardization. (2010). Environmental management systems -- requirements with guidance for use. Retrieved from http://www.iso.org/iso/catalogue_detail?csnumber=31807

[20] Kausek, J. (2007). Ohsas 18001: Designing and Implementing an Effective Health and Safety Management System. Government Institutes.

[21] ANSI/AIHA Z10 Committee. (2005). ANSI/AIHA Z10- 2005 Occupational Health and Safety Management Systems. AIHA.

[22] National Association of Environmental, Health & Safety Management. (2010). The National Association for Environmental, Health & Safety Management. Retrieved from http://www.naem.org/

[23] Gelfand, J.L. (2007, October 18). Flu Statistics: What are Your Odds of Getting the Flu?. Retrieved from http://www.webmd.com/a-to-z-guides/flu-statistics

[24] *Ibid.*

[25] Siciliano, G. (2003). Finance for Non-Financial Managers. USA: McGraw-Hill.

[26] Harvard Business School. (2002). Finance for Managers. Boston, MA: Harvard Business School Publishing Corporation.

[27] United States Department of Labor. (2010). Bureau of Labor Statistics. Retrieved from www.bls.gov/

[28] Wikipedia.org. (2010). Article on Bureau of Labor Statistics. Retrieved from http://en.wikipedia.org/wiki/Bureau_of_Labor_Statistics

[29] Council of State and Territorial Epidemiologists. (2010). Retrieved from http://www.cste.org/dnn/

[30] Wikipedia.org. (2010). Article on the Council of State and Territorial Epidemiologists. Retrieved from http://en.wikipedia.org/wiki/Council_of_State_and_Territorial_Epidemiologists

[31] Mountain and Plains Educational Research Center. (2010). Retrieved from http://maperc.ucdenver.edu/

[32] The Global Reporting Initiative. (2010). Retrieved from http://www.globalreporting.org/Home

[33] Wikipedia.org. (2010). Article on the Global Reporting Initiative. Retrieved from http://en.wikipedia.org/wiki/Global_Reporting_Initiative

[34] Puget Sound Chapter of the Human Factors and Ergonomics Society. (2010). Retrieved from http://www.pshfes.org/cba.htm

[35] The University of Texas Health Science Center at Houston. (2010). Safety, Health, Environment and Risk Management. Retrieved from http://www.uth.tmc.edu/safety/Annual%20Reports.html

[36] American Society for Healthcare Risk Management. (2010). *The Risk Management Handbook, 6th Ed.*

[37] International Organization for Standardization. (2008). Risk management-Principles and guidelines on implementation. Draft International Standard ISO/DIS 31000, (992), Retrieved from http://www.rmia.org.au/LinkClick.aspx?fileticket=AWkZuS%2BB6Wc%3D&tabid=85&mid=634

[38] American Industrial Hygiene Association. (2007). Industrial Hygiene & Safety Auditing - A Manual for Practice. Fairfax, VA.

[39] Jayjock, M.A., Lynch, J., & Nelson, D.I. (2000). Risk Assessment Principles for the Industrial Hygienist. American Industrial Hygiene Association. USA.

[40] Google. (n.d.). Definitions of Business Continuity on the Web. Retrieved from http://www.google.com/search?hl=en&rlz=1W1GGIE_en&defl=en&q=define:Business+continuity&sa=X&ei=Mx54TIXoB4i2sAOo3ISuBQ&ved=0CBkQkAE

[41] BSI Management Systems. (n.d.). BSI Standards and Publications: Business Continuity Management. Retrieved from http://orders.ceem.com/showitemlist.asp?category=78

[42] Sandman, P.M., & Lanard, J. (Instructors). Crisis communication: guidelines for action. [DVD]. Retrieved from http://www.aiha.org/education/dl/Pages/CrisisCommunicationGuidelinesforAction.aspx

[43] Woodcock, H.C., & Seibert, J. (2000). Investigations: A Handbook for Prevention Professionals. Fairfax, VA: American Industrial Hygiene Association.

[44] Cahill, L., & Kane, R.W. (2010). Environmental Health and Safety Audits. Government Institutes.

[45] Hiles, A. (2003). Business Continuity: Best Practices--World-Class Business Continuity Management. Brookfield, CT: Rothstein Associates Inc.

[46] Fulmer, K.L. (2004). Business Continuity Planning: A Step-by-Step Guide with Planning Forms on CD-ROM. Brookfield, CT: Rothstein Associates Inc.

[47] Hiles, A. (2007). The Definitive Handbook of Business Continuity Management. West Sussex, England: John Wiley & Sons Ltd.